Environmental Justice

Environmental justice has increasingly become part of the language of environmental activism, political debate, academic research and policy making around the world. It raises questions about how the environment impacts on different people's lives. Does pollution follow the poor? Are some communities far more vulnerable to the impacts of flooding or climate change than others? Are the benefits of access to greenspace for all, or only for some? Do powerful voices dominate environmental decisions to the exclusion of others?

This book focuses on such questions and the complexities involved in answering them. It explores the diversity of ways in which environmental and social difference are intertwined, and how the justice of their interrelationship matters. It has a distinctive international perspective, tracing how the discourse of environmental justice has moved around the world and across scales to include global concerns, and examining research, activism and policy development in the US, the UK, South Africa and other countries. The widening scope and diversity of what has been positioned within an environmental justice 'frame' is also reflected in chapters that focus on waste, air quality, flooding, urban greenspace and climate change. In each case, the basis for evidence of inequalities in impacts, vulnerabilities and responsibilities is examined, asking questions about the knowledge that is produced, the assumptions involved and the concepts of justice that are being deployed in both academic and political contexts.

Environmental Justice offers a wide-ranging analysis of this rapidly evolving field, with compelling examples of the processes involved in producing inequalities and the challenges faced in advancing the interests of the disadvantaged. It provides a critical framework for understanding environmental justice in various spatial and political contexts, and will be of interest to those studying Environmental Studies, Geography, Politics and Sociology.

Gordon Walker is Chair of Environment, Risk and Justice in the Lancaster Environment Centre at Lancaster University, UK.

D1270205

Environmental Justice
Concepts, evidence and politics

Gordon Walker

Routledge
Taylor & Francis Group

LONDON AND NEW YORK

First published 2012
by Routledge
2 Park Square, Milton Park, Abingdon, Oxon, OX14 4RN

Simultaneously published in the USA and Canada
by Routledge
711 Third Avenue, New York, NY 10017

Routledge is an imprint of the Taylor & Francis Group, an informa business

British Library Cataloguing in Publication Data
A catalogue record for this book is available from the British Library

Library of Congress Cataloging-in-Publication Data
Walker, Gordon.
Environmental justice / Gordon Walker.
p. cm.
Includes bibliographical references and index.
1. Environmental justice. 2. Environmental policy. 3. Environmental law.
I. Title.
GE220.W35 2011
363.7–dc22

2011016345

ISBN: 978-0-415-58973-4 (hbk)
ISBN: 978-0-415-58974-1 (pbk)
ISBN: 978-0-203-61067-1 (ebk)

Typeset in Times New Roman
by Cenveo Publisher Services

Contents

List of figures viii
List of tables x
List of boxes xi
Preface xii
Acknowledgements xiv

1 Understanding environmental justice 1
The scope of environmental justice 2
Framing 4
Claim-making 5
Definitions of environmental justice and the case for multiplicity 8
Defining environmental inequality: the is–ought distinction 12
Summary 14
Structure of the book 14
Further reading 15

2 Globalising and framing environmental justice 16
The environmental justice movement in the US 17
The international travelling of the environmental justice frame 23
Environmental justice framings of global issues 34
The implications of 'going global' 36
Summary 37
Further reading 38

3 Making claims: justice, evidence and process 39
The three elements of claim-making 40
Justice concepts: how things ought to be 42
Evidence: how things are 53
Process: why things are how they are 64

Summary 75
Further reading 75

4   Locating waste: siting and the politics of dumping   **77**
Resisting waste: three cases 78
Unequal patterns of waste site locations 84
Environmental racism or markets? Analysing positions 90
Displacement, toxic imperialism and environmental blackmail 94
From redistribution to prevention 100
Summary 101
Further reading 102

5   Breathing unequally: air quality and inequality   **104**
Air quality and multidimensional claim-making 104
Evidence of air quality inequality 107
Explaining patterns of inequality 111
Vulnerability and impacts on health: the 'triple jeopardy' 115
The distribution of responsibility for air pollution 118
Justice in the air 119
Summary 125
Further reading 126

6   Flood vulnerability: uneven risk and the injustice of disaster   **127**
Characterising flooding: values, time and nature 128
Inequalities in flood exposure: who lives with flood risk and why? 130
Inequalities in vulnerability: who suffers flood impacts? 135
New Orleans and the Katrina flood 139
Justice and flooding 148
Summary 154
Further reading 155

7   Urban greenspace: distributing an environmental good   **156**
Greenspace as a 'good thing' 157
Always a good thing? Contested meanings of urban greenspace 160
Greenspace and social difference: evidence claims and inequality 163
Greenspace and justice 173
Summary 177
Further reading 178

8   Climate justice: scaling the politics of the future   **179**
Challenges for climate justice 182

Impacts, vulnerabilities and adaptation 186
Mitigation, responsibilities and transitions 198
Towards integration in climate justice 209
Summary 212
Further reading 213

9   Analysing environmental justice: some conclusions   **214**
There is value in understanding exactly how social differentiation
  exists and how it is experienced in environmental terms 215
Environmental inequalities are constituted by more than spatial
  patterns of proximity and exposure 216
Recognising the methodological complexities and choices
  involved in generating empirical evidence is important in
  progressing understanding and in doing justice 216
It is necessary to distinguish between inequality and injustice
  and to reason carefully about why an inequality matters
  and to whom 217
Environmental justice is about more than just patterns of
  distribution; procedure, recognition and their detail,
  also matter 218
Environmental justice is contested and involves
  political challenges 219
Environmental justice is an objective but also a process
  of 'working towards' 220

Bibliography   *222*
Index   *249*

Figures

1.1 A cake fairly divided? 7
2.1 The globalising of environmental justice in two dimensions 17
2.2 Protest march by the Little Village Environmental Justice
Organization in Chicago 22
2.3 Friends of the Earth Scotland environmental justice
campaign leaflet, 2001 29
3.1 The three elements of environmental justice
claim-making 40
3.2 Explanatory interrelations between distribution, participation
and recognition 65
3.3 The framework of political ecology 71
4.1 The nuclear waste facility on the south side of Orchid Island,
Taiwan 81
4.2 Percentage of people of colour living near to hazardous waste
facilities in the US (1990 census data) 88
5.1 Environmental justice campaigning on air quality 105
5.2 Distribution of nitrogen dioxide by deprivation in England, 2001 112
5.3 Distribution of nitrogen dioxide by deprivation in Wales, 2001 113
5.4 Distribution of ward mean nitrogen dioxide exceedences for
England, 2001 121
5.5 Childhood asthma hospitalization patterns and transport and
waste facilities in Manhattan 124
6.1 Correlation coefficients between percentage of black and mean
altitude in urban areas of the US South 132
6.2 Percentage of total population within flood risk zones for
river flooding by deprivation decile 133
6.3 Percentage of total population within flood risk zones
for sea flooding by deprivation decile 134
6.4 An integrated view of disasters 136
6.5 Post-flood caravan living in Hull 137
6.6 Inundated areas in New Orleans following breaking of the
levees during Hurricane Katrina. New Orleans, Louisiana,
11 September 2005 140

6.7	Poverty and child poverty rates and percentage of African-Americans for areas across the region affected by Hurricane Katrina compared to national data	141
6.8	Comparisons of the five worst damaged New Orleans neighbourhoods	142
7.1	An urban community garden providing for social space and food production in New York	159
7.2	Urban greenspace in north London	161
7.3	A model of the elements shaping park use	162
7.4	Example of 600 metre 'buffers' around woodland in Scotland	167
7.5	Percentage of the Scottish population within 600 metres of new woodland by rural and urban deprivation decile	168
7.6	Harlem Piers water-front park in New York	175
8.1	Climate justice march in London	180
8.2	The complex coastline and fluvial geography of Bangladesh	189
8.3	Per capita emissions of CO_2 in 2006 for selected countries	199
8.4	Total emissions of CO_2 in 2006 for selected countries	200
8.5	Contraction and convergence of greenhouse gas emissions	203
8.6	Transport mode breakdown of average greenhouse gas emissions per person by income group, disaggregated by CO_2 and non-CO_2 emissions	205

Tables

1.1   The social and environmental dimensions of recent environmental
      justice research   2
2.1   Countries included in written material using an environmental
      justice frame   24
3.1   Issues in defining the appropriate metric of distributive justice
      for each of the topics of later chapters   43
3.2   Criteria for evaluating the degree of democracy of a siting process
      for a hazardous waste facility   49
3.3   Methodological complexities in GIS-based environmental
      inequality studies   58
4.1   Forms of injustice articulated in each of the three cases   83
4.2   Examples of longitudinal studies of waste site locations in
      the US and their key findings   90
4.3   Three different positions on environmental racism and waste
      facility siting and their evidence, justice and process components   91
5.1   An example of a set of claims made about air quality and its
      relation to poverty   106
5.2   Choices and complexities in the distributional analysis of air
      quality/pollution   108
5.3   Examples of regional and city scale analysis of the social
      distribution of air pollution in the US   110
6.1   Age and race profile of Katrina's victims in New Orleans   143
6.2   Concepts of justice in the Katrina flood   148
7.1   Types of open space   166
7.2   Proximity to greenspace for the top and bottom 10 per cent
      of income data zones in Glasgow   169
8.1   Ethical considerations in selecting adaptation strategies   195
8.2   Four different approaches to determining greenhouse gas
      emissions targets   201
8.3   Examples of potential distributional justice issues arising
      from carbon mitigation measures   208
8.4   Rights-centred principles for climate policy   210

Boxes

1.1 Examples of environmental justice claim-making 6
1.2 Some alternative definitions of environmental justice 8
1.3 Three concepts of justice 10
2.1 Executive Order 12898 and its troubled implementation 19
2.2 Mission statement of the Little Village Environmental Justice
Organization 21
2.3 Key aspects of the Environment Agency's position statement
on environmental inequalities (Environment Agency for
England and Wales 2004) 27
3.1 Alternative principles of justice in priority setting for
stream rehabilitation 45
3.2 River water quality and deprivation: a problematic research design 58
3.3 Environmental justice studies analysing cumulative risks 60
3.4 The urban political ecology manifesto 73
4.1 Waste and analytical themes 78
4.2 The two early studies of waste facility locations in the US 85
4.3 Toxic Wastes and Race at Twenty 87
4.4 Quantitative analysis of waste site locations in the UK 93
4.5 Trafigura and toxic waste dumping in Africa 96
5.1 Air quality and cross-cutting themes 105
5.2 Changing air pollution inequalities in Gary, Indiana 115
6.1 Flooding and cross-cutting themes 129
6.2 The 'forgotten city' of Hull and the 2007 flood 137
6.3 The Katrina Bill of Rights 152
7.1 Greenspace and cross-cutting themes 157
7.2 Children and greenspaces 164
8.1 Examples of how justice figures in positions on climate change
negotiations 181
8.2 Climate justice and cross-cutting themes 182
8.3 Projected climate change impacts in Bangladesh 189
8.4 Who dies in heatwaves? 191
8.5 Alternative principles for justly distributing adaptation
funding in the UK 194
8.6 The EJCC '10 Principles of Just Climate Change Policies in the U.S.' 211

Preface

It was about 13 years ago when I first wrote a short piece about environmental justice, speculating as to whether the political agenda that had become so vibrant and significant in the US would travel across the Atlantic to the shores of the UK. Since then much has changed – not least the academic literature on environmental justice which has expanded exponentially and diversified in fascinating ways. The language and ideas of environmental justice have travelled beyond the US, and it has now become a term far more widely used, studied and debated. Within the UK I had a small role in helping this happen. I led a series of research projects on environmental inequalities for the Environment Agency and for Friends of the Earth, and became engaged in many workshops, seminars and networking initiatives on environmental justice, including running a UK-wide seminar series on environmental inequalities during 2007–8.

This experience led me to reflect at some length on what environmental justice was all about. What was empirical research into patterns of environmental inequality really trying to achieve and what did the justice in environmental justice actually mean? I read more, thought more, learnt more, taught various groups of students about environmental justice, and the outcome of all this is somehow gathered together in this book. My intentions in writing the book, as laid out in the book proposal, were as follows. It would be a distinctive and accessible text on environmental justice written by someone working outside of the US context; it would use varied empirical material drawn from a range of countries, reflecting the globalization of environmental justice; it would lay out a framework for understanding different elements of environmental justice claim-making, integrating theoretical with empirical material; it would, in particular, deal with evidence and justice concepts together; and in doing all this it would be systematic, analytical and provide a structured pathway through a rapidly expanding and diversifying literature.

Now having completed the writing, I hope I have kept true to at least some of these ambitions. The book may not quite have the global scope intended, concentrating at times largely on where I know best (the UK) and where most of the research literature has been written about (the US), but I have tried to draw, where possible, on the rapidly expanding range of research outputs from other parts of the world. It hasn't quite succeeded in presenting the all-encompassing analytical framework I at one point thought possible, but it goes some distance in this direction. And I have tried to produce an accessible and thought-provoking text which should help readers delve further into the ever expanding and diversifying world of environmental justice journals, papers, books, reports and websites. Accordingly, at the end of each chapter there are suggestions for further reading and a summary of the chapter's key points and arguments.

I have had several audiences in mind. The book is written primarily for advanced undergraduate and postgraduate students, who may be studying a course/module on environmental justice (Geog 415 at Lancaster University), examining it as part of a broader programme of study, or undertaking a research project or dissertation. I have imagined these readers coming from a variety of disciplinary backgrounds – geography, sociology, politics, environmental studies/science, law, management and public health amongst others – and have therefore tried to write the book in a way that is broad rather than narrow in its conceptual assumptions and use of language. Undoubtedly there will be much that is inadequately covered from any one of these disciplinary perspectives, but there is, I firmly believe, value in the integration and broad coverage I have attempted. Non-academics, or, more correctly, those who are not currently involved in an academic institution, will also, I hope, engage with the material in this book and find it useful, for example in contextualising local cases of activism, understanding alternative ways of making claims about environmental justice, or assessing the scope for forms of policy innovation.

It may be helpful for you, the reader, to know something about me, the writer. I am a geographer and my disciplinary background inevitably shapes something of the content and preoccupations of the book. Consequently you will find plenty about spatial patterns and mapping; but there is also much about places, scales, processes and the politics of knowledge. Being a geographer gives me the scope to range freely across human–nature relations and to bring together a diversity of material from other disciplines. Geographers are often effective synthesisers and integrators, as I hope the scope of this book demonstrates, ranging as it does across multiple concepts, forms of evidence and sites of environmental justice politics.

Acknowledgements

In undertaking research and writing on environmental justice over the last ten or so years, I have worked with colleagues from many disciplines and backgrounds. I therefore owe thanks and many debts to a long list of people – beginning with Karen Bickerstaff, John Fairburn, Graham Smith and John Mooney, with whom I initiated an interest in environmental justice when at Staffordshire University, Gordon Mitchell at Leeds University, and a long list of colleagues from other universities with whom I collaborated on early research and writing projects: Kate Burningham, Jane Fielding, Sarah Damery, Judith Petts, Carolyn Stephens, Ruth Willis, Karen Lucas, Mark Poustie and others. Julian Agyeman has been particularly influential in shaping my understanding of environmental justice in the US and, along with Bob Evans, Sue Buckingham, Harriet Bulkeley and others in the Planning and Environment Research Group of the RGS-IBG, opened up a new agenda for UK environmental justice research. Malcolm Eames was also an excellent collaborator on the environmental inequalities seminar series and on other initiatives involving the Sustainable Development Research Network that promoted environmental justice ideas in government. Policy engagement was promoted by Helen Chalmers and John Colvin at the Environment Agency, whilst Simon Bullock and Duncan McClaren in varieties of Friends of the Earth gave me insights into the world of environmental justice related activism, as did Maria Adebowale from Capacity Global.

Learning about justice theory has been particularly challenging, and I patently still do not understand enough. But reading and listening to Andrew Dobson, Derek Bell, John O'Neill, Andrew Sayer and in particular David Schlosberg have helped me enormously. In the US, Ryan Holifield and Mike Porter helped to push on my conceptual thinking and were great joint conveners for an Association of American Geographers conference session in Chicago. Regular field trips with students to New York have been greatly enlivened by Cecil D. Corbin-Mark from West Harlem Environmental Action, who generously gave his time to inform them and me about the group's work and the 'toxics and treasures' of Harlem.

Writing this book began in Barcelona and finished in New York, with most of it completed in the slightly less glamorous locations of Leek, Keele and Lancaster (or on a train to or from Manchester). Music playing and listening did a lot to keep me sane, Todd especially. Arsenal (see Figure 7.2) excited and infuriated me in equal measure. Throughout the writing process I have been grateful to many people for their encouragement and advice: Fiona Tweed, Hugh Deeming, Noel Cass, Sam Brown, Betsy Olson, James Faulconbridge, Will Medd, Saskia Vermeylen, Maria Kaika and Elizabeth Shove; Anna Sera-Llobet and Nuria Calzada in Barcelona; Andrew Mould and Faye Leerink at Routledge; and the reviewers of the book proposal and finished draft. I am very grateful to Simon Chew at the Lancaster Environment Centre for producing most of the original and adapted figures in the book.

Permissions have been kindly given for the re-use and reproduction of various figures and images. An earlier version of Chapter 2 was first published by Sage Publications Ltd in *Global Environmental Politics* (2009), vol. 9, no. 3. Figure 2.2 is reproduced with the permission of Toban Black, Figure 4.1 with the permission of Chris Stowers, Figures 5.1 and 5.5 with the permission of West Harlem Environmental Action, Figure 6.5 with the permission of Rebecca Whittle, Figure 6.1 with the permission of the American Geographical Society and Figure 8.5 with the permission of the Global Commons Institute.

Finally, a very big thank you to Louise (including for reading through draft chapters), Sam, Hannah and Amy, who, having tolerated endless thumping of the laptop keyboard, will hopefully hear no more of that, for book purposes at least. The book is dedicated to them, as well as to my much missed Dad.

Note on the cover photo

The cover photo is a helium-powered banner, deployed by the **Backbone Campaign** near a now closed Detroit garbage incinerator. It was a solidarity action during the 2010 US Social Forum to support the work of the Zero Waste Detroit coalition.

The Backbone Campaign is US-based progressive movement-building organization that provides strategic tools and training to grassroots activists and organizations. Backbone specializes in "Artful Activism," i.e. the strategic use of creative tactics such as giant spectacle imagery, banners, puppets, flash mobs, and other forms of non-violent direct action. Some other examples of the above include the "Target Ain't People" flash mob, the widely photographed "Bush Chain Gang," and the 150-foot "We the People" Constitution Preamble action at the Lincoln Memorial.

Backbone Campaign was founded by artist-activists on Vashon Island, WA (near Seattle) in 2003. It evolved out of a local affinity group that formed in response to the regressive policies the G.W. Bush Administration. The organiza-tion's early projects such as Backbone Awards, Spineless Citations, and a 70-foot-long spine puppet delivered to the Democratic Party's Convention and headquarters in 2004 and 2005 were framed as "giving the Democrats a back-bone." Actually, fiercely independent of party politics, Backbone swiftly transi-tioned to an "emboldening citizens and leaders," frame using the backbone metaphor to symbolize "an interlocking agenda, a coalition, and the personal courage necessary to fight for a future worthy of our children."

Co-founder and Executive Director Bill Moyer, identifies himself as a "radical solutionary," and the organization's ideological leanings as progressive populist. Their motto is "When the People lead, the Leaders Follow."

Learn more at www.BackboneCampaign.org

1 Understanding environmental justice

This book is about the intertwining of environment and social difference – how for some people and some social groups the environment is an intrinsic part of living a 'good life' of prosperity, health and well-being, while for others the environment is a source of threat and risk, and access to resources such as energy, water and greenspace is limited or curtailed. It is also about how some of us consume key environmental resources at the expense of others, often in distant places, and about how the power to effect change and influence environmental decision-making is unequally distributed. Most fundamentally, it is about the way that people should be treated, the way the world should be.

The term that best captures this set of concerns is *environmental justice*. These two words have become used in many different ways – as a campaigning slogan, as a description of a field of academic research, as a policy principle, as an agenda and as a name given to a political movement. Emerging from its origins in anti-toxics and civil rights activism in the US to produce what some have seen as one of the most significant developments in contemporary environmentalism, environmental justice has become increasingly used as part of the language of environmental campaigning, political debate, academic research and policy-making around the world. As we shall see, we can now find examples of environmental justice language being used in countries as diverse as South Africa, Taiwan, Israel, Germany, Australia, Brazil and Scotland, and with reference to issues from the local street level through to the global scale. It has, as Agyeman and Evans (2004) argue, provided a 'vocabulary of political opportunity' and an important way of bringing attention to previously neglected or overlooked patterns of inequality which can matter deeply to people's health, well-being and quality of life.

This, as part of the discourse of contemporary political life, makes environmental justice significant and worthy of attention. More fundamentally, though, focusing on environmental justice provides a route into examining important aspects of how people think, reason and act in relation to environmental concerns. Justice *does* and *should* matter, as much to our environmental concerns and experiences as to others. And as we shall see, working out exactly how justice or fairness matters, and the parameters within which claims and judgements of environmental inequality and injustice can be made, provides just as much scope for deliberation and debate as more familiar and established justice concerns.

The scope of environmental justice

In this book I aim to explore the diversity of ways in which environment and social difference are intertwined and how the justice of their interrelationship matters. As environmental justice language has moved spatially around the world and across scales to include global concerns, so the scope of what has been positioned within an environmental justice 'frame' has expanded and diversified (Holifield *et al.* 2010; Sze and London 2008). In its early formulations in the US in the 1980s, environmental justice activism and research focused pretty narrowly on the relationship between race and poverty and the spatial distribution of waste and industrial sites producing pollution impacts, including accusations that a form of 'environmental racism' deliberately targeting poor black communities in locating polluting sites was at work (see discussion in Chapter 4). Whilst this is still an important and distinctive theme, over the ensuing 30 years far more has been encompassed.

A review in 2005 of the content of environmental justice activist group websites in the US identified 50 distinct and varied environmental themes (Benford 2005), including transport issues, food justice, deforestation, lead poisoning, bio-piracy and transportation. Looking to the research literature, a similarly expansive field of study is encountered (see Table 1.1). The forms of social difference that have been featured in recent environmental justice research

Table 1.1 The social and environmental dimensions of recent environmental justice research

Social dimensions	Environmental dimensions	
Race	Air pollution	Greenspace
Ethnicity	Accidental hazardous releases	Outdoor recreation
Class	Waste landfills	Mineral extraction
Income	Waste incinerators	Hog industry
Deprivation	Contaminated land	Emissions trading
Gender	Brownfield land	Oil drilling and
Single parent families	Urban dereliction	extraction
Households in social housing	Lead in paint and pipes	Access to healthy food
Older people	Flooding	Fuel poverty
Children	Noise	Wind farms
Indigenous peoples	Drinking water quality	Nuclear power stations
Disability	River water quality	Climate change
Deafness	Transport	Trade agreements
Special needs	Forest fires	Alcohol retail outlets
Future generations	Whaling	Biodiversity and genetic
	Wildlife reserves	resources
	Agriculture	Genomics
		Land reform

(the left-hand column in Table 1.1) include, for example, questions of age, the environmental rights of indigenous people, gender differences, the environmental and participatory concerns of disabled people and responsibilities to future generations. The range of environmental concerns that have featured in the environmental justice research literature (the right-hand side of Table 1.1) is now vast – from landfills to oil extraction, lead in paint to whaling, wind farms to hog farms – and covers a wide diversity of environmental risks, benefits and resources.

In later chapters we will examine a selection of these environmental concerns in some detail – waste, air pollution, flooding, greenspace and climate change. In each of these we will consider some of the evidence of unequal patterns and experiences for different social groups, and the arguments and claims that have been made in research and in environmental campaigning. Some of this evidence and argumentation is striking and compelling, and some enormously important work has been undertaken over recent years to show how environmental inequalities are experienced, how they are caused and how people's living conditions, access to environmental resources and access to basic democratic rights need to be addressed. However, such material will not be presented uncritically, and throughout the book I am hoping that readers will be encouraged to think about what is being asserted and argued and to develop their own critical evaluations as to its meaning and importance.

In this vein, the book has a broader aim of developing analytical insight and understanding in an academic field that sometimes lacks a more critical edge. This is to be achieved by tracing the growth, spread and evolution of environmental justice activity over recent decades, examining some of the vast range of evidence, arguments, explanations and demands that have been put forward, and considering the implications that then follow. Some key questions underpin this endeavour:

- Is there one definition of what constitutes environmental justice (and injustice) or are there many potential different ones? As the language of environmental justice has evolved from its origins in the US and become used in many different places, contexts and circumstances, what does this imply for what environmental justice is taken to mean?
- How can we pick our way through the many types of environmental inequalities and forms of justice and injustice now being examined around the world and work out a way of categorising, comparing and evaluating what is at issue?
- What are the methods through which evidence of environmental inequalities is being produced and what are the complexities involved in applying these methods and making sense of the evidence?
- Are there ways in which we can analyse the evidence or 'knowledge claims' being made by an environmental group, a scientist or a local resident and understand why evidence is disputed and disagreements erupt?
- What alternative explanations are there of the processes that have produced and sustained patterns of inequality and injustice in different contexts?

These questions all encourage an analytical take on the meaning of environmental justice and require tools for the critique and evaluation of what is being argued for and about. In the rest of this chapter the first steps towards developing this approach and towards answering some of these questions will be laid out. Two key ideas will first be discussed – framing and claim-making – before focusing on how environmental justice can be defined and understood. Through this discussion I introduce some important ideas for the rest of the book.

Framing

The notion of an 'environmental justice frame' has already been referred to and will be a recurrent reference point throughout the book. Concepts of frames and framing have taken root in various areas of social science but have been particularly powerful in the analysis of social movements or 'collective action', including that of the environmental justice movement (Capek 1993; Faber 2008; Taylor 2000). Framing is a notion that recognises that the world is not just 'out there' waiting to be unproblematically discovered, but has to be given meaning, labelled and categorised, and interpreted through ideas, propositions and assertions about how things are and how they ought to be. By implication there is not just one interpretation of the world available, but alternative versions, multiple versions (the many alternative religions are an obvious general example). Applying this multiplicity to environmental justice concerns, we can see how a pattern of environmental inequality might be interpreted as 'just how things normally are', as the outcome of how the market economy works, or as the result of systematic discrimination and injustice. A 'problematic' environmental risk may be interpreted as something to be managed through good science, or as the consequence of the capitalist pursuit of profit by some at the expense of others, or, indeed, as not a problem at all.

Social movements, such as the environmental justice movement, actively try to persuade others of their preferred frames of meaning, interpreting what is wrong with the world and advocating change (Benford and Snow 2000). Some of these framings are quite radical in making a case for a different way of organising society and addressing environmental concerns. But they are not alone in this endeavour. Others engage in their own work of framing – governments and political parties do it all the time, as do the media and corporate actors. Frames are contested and argued about and counter-frames are deployed to challenge dominant or threatening alternatives. An example is provided by Shibley and Prosterman (1998) in their analysis of competing frames in media coverage of childhood lead poisoning. They trace the difficulties environmental justice activists have in establishing a framing of lead poisoning as a threat to health that is particularly acute for some children in US society, rather than as a 'silent epidemic' that is a risk to all children which stands as the dominant frame.

Academics engage in framing as well. There is an academic frame of work on environmental justice that I am writing within, which has certain shared ideas, terms and conventions, even if I might be trying to stretch and interpret these in particular ways. What is interesting about the frames that come to be is where they have come from, what they include and leave out, and what difference they make. Also of interest is how they appear in different forms in different places and how they evolve and become more or less powerful and relevant in the processes that they themselves are part of. At various points in this book I will be asking these questions, not only in Chapter 2, which is most directly concerned with tracing the appearance and evolution of the environmental justice frame within political activity around the world, but also in other chapters as particular topics, approaches and contexts are considered.

Claim-making

A second term that will be widely used throughout the book is *claim-making*.[1] This, like framing, is used to emphasise that there are many different ways in which we can try to make sense of, or make claims about, the world around us. As will be discussed in Chapter 3, claims about environmental justice can have different elements or components to them, and analytically we can identify these and categorise and evaluate them. For example, and to draw on the topic of Chapter 5, a claim about the justice of distribution of air quality in a city might involve:

- claims about concentrations of air pollutants and how these are concentrated in particular parts of the city;
- claims about the vulnerability of old, young or poor people to the health effects of polluted air;
- claims about responsibility for the production of the poor air quality;
- claims about why the distribution of poor air quality is unjust or unfair;
- claims about what would constitute a just or fair way of addressing this situation.

Such elements of claim-making are open to further analysis to bring out, for example, how they are drawing on particular types of quantitative or qualitative evidence, particular concepts of justice or particular notions of responsibility. There is too much here to cover fully at this point, but the different possibilities are important, as are the ways different elements are combined in claim-making and how these combinations might become more or less effective in achieving the aims of the actors involved.

One of the basic combinations that is often made within justice claim-making is to link evidence of a condition of inequality with a normative position on what is just or unjust (see the later discussion of the distinction between inequality and injustice). Box 1.1 shows four examples of such combinations where in each case descriptive evidence of a situation (in these cases the distribution of waste sites,

---

1 I have deliberately chosen to use 'claim-making' rather than 'claims-making'. The latter plural version is often used within sociology, but no significant distinction is implied.

Box 1.1 Examples of environmental justice claim-making

'There is a disproportionate concentration of landfill waste sites in communities with a high proportion of African-American people and this is wrong,
unfair and racist.'

'Access to a park or to a green area shouldn't be only for people who are
fit and healthy, its not only them that need nice places to go; but around
here nothing is done to enable access for people who are disabled or who
have problems with getting around.'

'We were obstructed from getting hold of information on levels of contamination and that's wrong, everyone should have access to information about
threats to their health.'

'The richer countries of the world have produced most of the carbon
emissions contributing to climate change, and it is not fair that it is the
poorest countries and the poorest people in those countries that will suffer
the worst consequences.'

---

access to parks, access to information, carbon emissions and climate change
impacts) is linked directly to a normative claim about what is just or fair.

The point here is that the academic literature on environmental justice has
tended to focus either on analysing justice concepts and theories – drawing on
various philosophical and political traditions (Dobson 1988; Schlosberg 2007;
Wenz 1988) – or on the generation of evidence of patterns of inequality. Rarely
have the linkages been adequately explored or both elements been approached as
forms of claim-making. One of the objectives of this book is therefore to explore
the possibilities of doing this and of taking a more integrative approach.

What the examples in Box 1.1 also demonstrate is that acts of justice claim-
making are essentially open to all. You and I (and my children in particular)
routinely make claims about the justice or injustice of a situation. It is commonplace and seemingly inherent to being a social being. As Sayer (2005: 5) states,
'in everyday life the most important questions tend to be normative ones'. My
children have always made claims (we would call it arguing) about how fairly
they are treated, about whether or not their brother or twin sister got more birth-
day cake than they did (Figure 1.1), about how many times they got to sit in the
front seat of the car and so on (familiar to all parents and siblings I'm sure). They
could also make such claims about environmental conditions, about their capacity
to have an influence on decisions that affect their environment and the environment of others, and about the way that the consumption practices of our society
are having serious impacts on other people elsewhere in the world. As they grow
up, I hope that they do articulate and make such claims, as they see fit. In other
words, there is an everyday voice and mode of claim-making that needs to be

Figure 1.1 A cake fairly divided?
    Source: The author.

incorporated into our understanding of justice, environment and social difference (not just the voices and modes of the politically engaged) and in this book at various points I will endeavour to bring this into view.

This point has a corollary: that claims made by those with particular professionalised roles and expertise must similarly also be seen as claims, rather than assertions of absolute truth based on their 'better', 'more expert' grasp of what is at stake. An analysis that I (an academic with qualifications and letters after my name to prove it) might produce of the relationship between patterns of air pollution and patterns of social deprivation (see Chapter 5) might be grounded in data and statistical methods, but clearly there are sufficient assumptions, uncertainties and unreliabilities in any such analysis to make its conclusions provisional and contingent rather than definitive. Claims rather than truths. You may decide you are perfectly happy with the analysis I have undertaken and convinced by the assertions I am making – convinced that it is a better claim about patterns of inequality than others might be, because it is backed up by good enough evidence and it makes a reasoned case; but I would rather not take that for granted. It then becomes interesting to think about on what grounds, in what circumstances and for what reasons some claims are advocated and given more authority and respect than others.

As we shall see in later chapters, disputes can open up about both what constitutes reliable evidence and the degree to which injustice of some form can be 'proven' to exist. These are not usually disputes that can simply be resolved by collecting better evidence or doing better analysis, as politics and ideology are also typically (if not always) at work. Hence arguments in favour of greater reliance on strictly applied scientific methods to establish patterns of exposure to risks (Bowen and Wells 2002) can be made just as strongly as claims that science has been corrupted by state and corporate interests to hide and deny patterns of harm amongst vulnerable communities (Faber 2008).

Definitions of environmental justice and the case for multiplicity

Having introduced the framing and claim-making that are involved in environmental justice discourse, what then follows for how we understand the defining of what environmental justice is? Looking across academic, activist and policy literatures, environmental justice is most often defined in terms of an *objective*, something that is sought after and for which certain conditions are specified. The act of producing and publicising an objective-based definition is a key part of constructing a politically powerful environmental justice frame around which people are to be recruited and mobilised. An objective also does the important job of providing a metric or standard against which current conditions can be judged and critiqued, and from which claims can then be constructed.

From what has already been said we might well expect that people could have different ideas about how to define environmental justice as an objective. And yes, when we look across academic, activist and policy literatures, we do not readily find one agreed definition of environmental justice being used, but rather multiple alternatives. To illustrate this, Box 1.2 reproduces six definitions of environmental justice taken from a range of different sources, places and contexts.

Box 1.2 Some alternative definitions of environmental justice

Commonwealth of Massachusetts

'Environmental justice is the equal protection and meaningful involvement of all people with respect to the development, implementation and enforcement of environmental laws, regulations and policies and equitable distribution of environmental benefits.' (Commonwealth of Massachusetts 2002: 2)

US Environmental Protection Agency (EPA)

'Environmental Justice is the fair treatment and meaningful involvement of all people regardless of race, color, national origin, or income with respect to the development, implementation, and enforcement of environmental laws, regulations, and policies ... It will be achieved when everyone enjoys the same degree of protection from environmental and health hazards and equal access to the decision-making process to have a healthy environment in which to live, learn, and work.' (US Environmental Protection Agency 2008)

Friends of the Earth Scotland

'Environmental justice is the idea that everyone has the right to a decent environment and a fair share of the Earth's resources.' (Friends of the Earth Scotland 1999)

Coalition for Environmental Justice (in Central and Eastern Europe)

'A condition of environmental justice exists when environmental risks, hazards, investments and benefits are equally distributed without direct or indirect discrimination at all jurisdictional levels and when access to environmental investments, benefits, and natural resources are equally distributed; and when access to information, participation in decision-making, and access to justice in environment-related matters are enjoyed by all.' (Steger 2007)

Bunyan Bryant

'Environmental justice refers to those cultural norms, values, rules, regulations, behaviours, policies and decisions to support sustainable communities, where people can interact with confidence that their environment is safe, nurturing and productive. Environmental justice is served when people can realize their highest potential, without experiencing the 'isms'. Environmental justice is supported by decent paying and safe jobs, quality schools and recreation; decent housing and adequate health care; democratic decision making and personal empowerment; and communities free of violence, drugs and poverty.' (Bryant 1995a: 6)

Carolyn Stephens, Simon Bullock and Alistair Scott

'Environmental justice means that everyone should have the right and be able to live in a healthy environment, with access to enough environmental resources for a healthy life; that responsibilities are on this current generation to ensure a healthy environment exists for future generations, and on countries, organisations and individuals in this generation to ensure that development does not create environmental problems or distribute environmental resources in ways which damage other people's health.' (Stephens *et al.* 2001: 3)

Considering the words contained in these six definitions with some care is an instructive exercise. All of the definitions are concerned with justice to 'people'. This is a key common and distinguishing feature that separates environmental justice from notions of ecological justice or justice to non-humans (Low and Gleeson 1998; Schlosberg 2007). However, the particular ways in which the populace is divided up into groups varies. For several definitions it's just 'everyone', but for the US EPA it's 'race, color, national origin, or income' that particularly matter. For Stephens *et al.* (2001) 'future generations' are important, but our children and grandchildren do not appear specifically in any other definition. Nowhere are other particular social categories, such as gender or age, highlighted.

In terms of the ways in which the environment matters, for Bunyan Bryant it is a range of dimensions including that it is 'safe, nurturing and productive', whilst for the Commonwealth of Massachusetts it is 'the distribution of environmental benefits' and for the Coalition for Environmental Justice the four dimensions of 'environmental risks, hazards, investments and benefits'. The only definitions to consider issues of consumption and responsibilities to others are Scottish Friends of the Earth in its 'fair share of the Earth's resources' and Stephens *et al.*'s injunction that we do not 'distribute environmental resources in ways which damage other people's health'.

In terms of basic concepts of justice – summarised in Box 1.3 and discussed in much more detail in Chapter 3 – all of the definitions include notions of distributive justice, who lives with, consumes or receives what. Most, but not all, also tackle questions of procedural justice: 'access to information, participation in decision-making, and access to justice' for the Coalition for Environmental Justice, and 'equal access to the decision-making process' for the US EPA. Bunyan Bryant's definition makes reference to the 'isms' and to 'cultural norms and values', which are expressions of justice as recognition.

Looking across all of these definitions, we find some commonalities but also much diversity in what exactly is at stake and what environmental justice (and

---

Box 1.3 Three concepts of justice

Distributive justice – justice is conceived in terms of the distribution or sharing out of goods (resources) and bads (harm and risk)

Procedural justice – justice is conceived in terms of the ways in which decisions are made, who is involved and has influence

Justice as recognition – justice is conceived in terms of who is given respect and who is and isn't valued

See Chapter 3 for fuller explanations and discussion.

injustice) is taken to mean. If we also pick on particular conditioning terms such as 'meaningful involvement', 'interact with confidence', 'a healthy life', 'without direct or indirect discrimination' and 'fair share', we could ponder and argue at length about what each of these means and how they might be operationalised. Although this is not quite the point – because these are general statements and principles and it is facile to criticise them for not being precise enough and for being 'open to interpretation'. However, this exercise does demonstrate the problems we might well have in trying to settle on one unified and agreed definition and vision of what environmental justice is, and how we would know that we had it. More fundamentally, it is argued in Chapter 2 that environmental justice is situated and contextual, grounded in the circumstances of time and place, hence defying universal definition – although common and recurrent elements do exist, as evident across the Table 1.2 definitions.

Such a perspective has become more common in the literature on environmental justice, although some, such as Ikeme (2003), have made appeals for greater conceptual clarity and precision and for the adoption of a 'unifying framework'. Wenz (1988: 2) made one of the first cases for a plural understanding of environmental justice, arguing that different perspectives on justice can often be found to underlie environmental disputes:

> disputes about injustice are common. Many of these disputes are fostered by differing conceptions of justice. Because people have different ideas about justice, a social arrangement or environmental policy that one person considers just will be considered unjust by another.

Phillips and Sexton (1999: 2) also see that there are many 'legitimate' definitions and that significant consequences flow from the choices that are made:

> there are many possible legitimate definitions depending on one's beliefs, opinions, and values. The central point is not that a particular definition is right or wrong, but rather that choosing a definition has distinct implications for the formulation, implementation, and evaluation of both policy and science.

However, it is David Schlosberg who has done most to convincingly and expertly lay out the grounds for what he calls a 'multivalent' understanding of environmental justice in both theory and praxis. He persuasively shows in his most recent work (Schlosberg 2007) how different concepts of justice are integrated in the arguments and discourses of environmental justice activists, both in the US and in global justice movements, and that they in this way accept 'both the ambiguity and the plurality that come with such a heterogeneous discourse' (ibid.: 5). Indeed, he argues that 'within the environmental justice movement, one simply cannot talk of one aspect of justice without it leading to another' (ibid.: 73). In Chapter 2 we will also see how the emergence of environmental justice ideas and frames around the world leads to an argument for a relative and contextualised understanding of what constitutes environmental justice, rather than one searching for universal meaning and conformity.

Such perspectives can be unsettling for those looking for simplicity and clarity, but it is both an honest reflection of 'what is out there' and a necessary step in developing a more sophisticated understanding of the meaning that has been given to environmental justice in different contexts and that *could* be given to it in the future. It is quite possible, as many already do, to select an environmental justice definition that is most satisfactory and work with it in concrete terms, to use it as a guide and to make arguments, if necessary, as to why it is better than other ones. The fact that others might have alternative definitions for the same two words is not then necessarily a problem, but may in fact present opportunities for dialogue and discussion – an argument widely made about 'sustainable development', which has even more of a 'problem' of multiple definitions (Walker and Shove 2007). Justice is inevitably political and politics involves disagreement, competing perspectives and active work to persuade others of your point of view.

Defining environmental inequality: the is–ought distinction

A final introductory step in setting up the analytical approach in this book is to consider the meaning of inequality. Having argued that we should expect multiple meanings to prevail, in particular around contested concepts such as justice and fairness, I am now going to be maybe rather perverse and try to draw a precise line in the sand between the notions of environmental inequality and environmental (in)justice. This is not because I demand that everyone just shapes up and becomes more accurate in how they use the term and that only one meaning will do. Rather, it is because I have found it useful, in being analytical, to make as clear as possible a distinction between inequality and injustice, particularly when considering the practices of claim-making.

So for me – and I hope as consistently applied in this book – inequality is a *descriptive* term, describing a condition of difference or unevenness of something (such as income, health, pollution exposure/creation, opportunity, influence, access to resources, consumption of resources), between different groups of people (old/young, black/white, rich/poor, north/south, this generation/future generation, etc.). Accordingly, inequality can be measured and described using data of various potential forms – although such description will never be an entirely neutral or unconstructed exercise.

In some research and policy domains the use of the term inequality or equality also carries normative qualities – inequality as something always negative and to be removed, equality as something always to be sought after. However, I have found it useful to resist such a move. As Harvey (1996: 5) argues, it is necessary to consider 'the just production of just geographical differences' if we are to make sense critically of the many, if not infinite, varieties of unevenness that do undoubtedly exist. Or as Wenz (2000: 175) puts it, 'equality is presumptively just but the presumption can be rebutted'. What is unequal will not be considered always and everywhere undesirable, bad, unfair or unjust. Some form of judgement or claim has to accompany this, for example, about the severity, consequences or morality of the inequality and the need for it to be reduced or removed. This separation of description and prescription, between 'is' and 'ought' (Proctor 2001), is an important distinction in much moral philosophy and helpful in being analytical, as this book is seeking to be.

Making this distinction is therefore important in promoting better reasoning about what constitutes environmental justice and injustice. In the geographical research community in particular (the discipline that I feel best able to engage with at this point) the bringing of ideas of justice more fairly and squarely into research and writing about the environment has been rather bereft of sustained reasoning about what the justice in environmental justice should constitute and why. Many geographers working within this framing have assumed that injustice is self-evident and unproblematic, that evidence of spatial-distributional inequality can be simply equated with injustice, that it is wrong in some way, without needing to explain for what reason(s). There are exceptions, some already having been referred to, but these are overwhelmed by the weight of largely 'uncritical' environmental justice scholarship that has either attempted to develop the 'facts' of unequal distribution of environmental 'goods' and 'bads' (values taken uncritically) or followed the resistance work of environmental justice activism without evaluating its normative foundation.

Indeed my own engagement with environmental justice research has not been without fault. Initially intent on reproducing for the UK the types of distributional studies that had been so influential in the emergence of environmental justice as a political force in the US (see Chapters 2 and 4), my concern was with geographical patterns of distribution of available environmental parameters (air quality, flood risk, greenspace and so on) and how these intersected with measures of social difference (see Fairburn *et al.* 2005; Walker *et al.* 2003, 2006). Some striking patterns were revealed (as discussed in Chapters 5, 6 and 7), but these were claimed patterns of difference and inequality, not directly of injustice. Whilst the various reports we produced acknowledged – in asides and recommendations – that questions of justice and fairness remained to be determined, this work was left for others to do. It soon became apparent, however, that without carefully reasoned accounts of the ways in which socio-environmental inequality mattered and 'injustice' was being produced, the value of revealing difference was severely diminished. How, for example, did poor river water quality actually matter to the predominantly poor urban communities who lived near to it, and how was their well-being diminished? Why should enabling proximity and access to greenspace for different social groups be a policy priority? Could a community surrounded by landfill waste sites in Scotland still be a case of 'environmental injustice' even though it was predominantly white and lower-middle class in social make-up? So in analytical terms being descriptive with inequality and normative with justice, and in this way maintaining the 'is–ought' distinction, is, sometimes at least, a productive thing to be.

Summary

In this opening chapter I have endeavoured to lay out an approach to making sense of environmental justice and explained how I intend to write about it in the rest of the book. We have seen how environmental justice at its broadest is about the intertwining of environment and social difference. We have seen how the field of environmental justice activism, research and policy has moved in all sorts of interesting ways geographically and into a wide diversity of forms of social difference and types of environmental concern – meaning that we are in complex and interesting rather than simple and obvious territory.

In order to handle this complexity and begin to understand what working with the language of environmental justice involves, I have introduced two connected concepts that are useful for developing a critical perspective. These are *framing*, the process of making sense of the world and putting forward and naming preferred ideas and meanings, an activity undertaken by environmental justice activists who try to enrol others into their campaigning or collective action frames (but also, in a less overtly political way, by researchers and other actors); and *claim-making*, the process of making various forms of claim about the conditions of a situation, such as a pattern of environmental inequality, and the extent to which this is just or unjust. Claim-making is typically multidimensional, involving various component elements that can be identified, categorised and interrelated.

We have seen how there is not just one environmental justice frame or one agreed definition of what a just environmental condition consists of, but rather multiple alternatives being applied in different contexts. I have argued that it is futile to expect that a single environmental justice can be found around which an absolute consensus can be constructed and that will happily serve in all circumstances, for all frames and instances of claim-making. Rather, I have aligned myself with those who argue for an openness to diversity in the different ways that environmental justice is understood and applied.

We have also seen that a range of different actors are involved in being concerned about justice and the environment – from the lay citizen to the activist and the 'expert' scientist or academic – and that we can position all forms of assertion made by these actors as claims, leading us to focus attention on what it is that makes some forms of evidence and justice claim more powerful, convincing and influential than others. I have also, in related terms, called for better critique and more active reasoning about what constitutes environmental injustice, something which I argue is helped by maintaining a distinction between inequality as descriptive and injustice as normative concepts.

Structure of the book

The next two chapters build on the ideas and themes of this first chapter by dealing, in Chapter 2, at greater length with the history, evolution and globalisation of the environmental justice frame and what this implies for how we understand the meaning of environmental justice, and by putting forward, in Chapter 3, a framework for examining the constituent elements of environmental justice claim-making. This framework distinguishes between the justice, evidence and process elements of claim-making, and through examining each of these elements Chapter 3 provides a wide-ranging suite of resources for the critical analysis of particular instances of environmental justice research, advocacy, discourse and policy.

The rest of the book consists of five chapters focused on environmental justice in relation to specific environmental topics or domains – wastes of various forms (Chapter 4), air pollution (Chapter 5), flooding (Chapter 6), greenspace (Chapter 7) and climate change (Chapter 8) – which explore, illustrate and explicate particular sets of the cross-cutting themes introduced in Chapters 1–3. The choice of these topics is intended to encompass well-established matters of concern (waste and air pollution) as well as those that have been more recently positioned within an environmental justice frame (flooding, greenspace and climate change), addressing local through to global scale concerns. There are many other topics that these chapters could have focused on, but they provide more than enough environmental and social variety to contend with in considering the complexities of evidence and the determination of what environmental justice constitutes in context.

In the final chapter a series of conclusions are drawn that emerge from and consolidate the learning that has been achieved through the book, and which advocate ways of taking forward the analysis of environmental justice in the future.

Further reading

To explore further the range and variety of environmental justice activism and scholarship browse the contents of the journals *Environmental Justice* (www. liebertpub.com/env) and *Local Environment: International Journal of Justice and Sustainability* (www.tandf.co.uk/journals). There are various edited books that are also a good starting point, including Agyeman *et al.* (2003) and Pellow and Brulle (2005).

Several websites have been set up to provide resources on environmental justice, including:

- Environmental Justice Resource Centre at Clark Atlanta University, www. ejrc.cau.edu/
- Center for Environmental Justice and Children's Health, www.nd. edu/~kshrader/cejch.html
- National Black Environmental Justice Network, http://www.nbejn.org/ who.html
- Environmental Justice Research and Resources (at Lancaster University), http://geography.lancs.ac.uk/EnvJustice/.

2 Globalising and framing environmental justice

The notion that environmental justice can be understood as a 'frame' was introduced in the previous chapter. Framing is a notion that recognises that the world is not just 'out there' waiting to be unproblematically discovered, but has to be given meaning, interpreted through ideas, propositions and assertions about how things are and how they ought to be. In this light, to talk of environmental justice is to suggest a particular way of making sense of the world, specifically of interpreting and evaluating the intertwining of environment and social difference (if we follow the very broad definition provided in the last chapter). In this chapter we will examine in some detail the characteristic elements that have been involved in environmental justice framing, in order to analyse how the use of an environmental justice frame has evolved, diffused and travelled around the world. The focus is on political activity, particularly by environmental justice advocates, activists and campaigning groups, but also encompassing the frames used by government bodies and state agencies. As already argued, the use of justice arguments in relation to socio-environmental concerns is not restricted either to activity that takes place under an 'environmental justice' label or only to actors operating overtly within political and public arenas. In this chapter, however, the remit is on environmental justice (those two words explicitly) as a political label, how this first materialised, the ideas and meanings it has conveyed and how these have evolved.

A narrative running through this chapter is the globalising of environmental justice. There are numerous excellent accounts of the origins, development, problems and successes of the environmental justice movement in the US (Bullard 1999; Faber 2008; Pellow and Brulle 2005; Schlosberg 1999; Shrader-Frechette 2002). It is not my intention to duplicate these accounts, beyond what is necessary to provide a good outline and a foundation for later comparative analysis. What is interesting beyond the core US experience, though, is how a range of alternative versions of the environmental justice frame have emerged in other parts of the world and how these have increasingly engaged with issues that cross national borders. This, as we shall see, has implications for how we understand environmental justice and also the dynamic, grounded geography of framing processes.

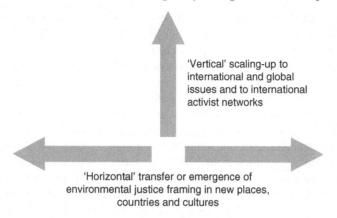

'Vertical' scaling-up to
international and global
issues and to international
activist networks

'Horizontal' transfer or emergence of
environmental justice framing in new places,
countries and cultures

Figure 2.1 The globalising of environmental justice in two dimensions.

This chapter uses a simple categorisation for analysing the globalisation of environmental justice in two related dimensions (see Figure 2.1). The first dimension involves the 'horizontal' emergence of the language and rhetoric of environmental justice in new settings around the world. Here the international utilisation of the frame can be mapped, along with analysis of the processes of diffusion, reproduction and contextualisation that have taken place within the political and institutional cultures of different countries. The second dimension of globalisation involves the 'vertical' extension of the scope of environmental justice frames to encompass concerns that do not end at national borders but which involve relations between countries and global scale issues. Through such 'scaling up', environmental justice activism can no longer be characterised as being only about local disputes or 'militant particularisms' (Harvey 1996) and the intranational distribution of environmental bads (Dobson 1988). This again has implications for how we understand the scope of environmental justice and the practice of activism from local to transnational scales.

It is impossible to take stock of each and every application of an environmental justice frame around the world, so I have selected the cases of the UK and South Africa for closer examination. However, it is with the original framing processes, the collective political action in the US that started it all, that we will begin.

The environmental justice movement in the US

The environmental justice movement in the US has been specifically analysed in terms of the framing work involved (Sandweiss 1998; Taylor 2000). Framing work within social movements has a number of typical elements (Benford and Snow 2000). These include the articulation of normative ideas (visions and objectives for how things should be), the diagnosis of problems and responsibilities for problems, and the prognosis of solutions and processes of change. Gamson (1992) also argues that notions of justice and injustice are routinely part of the framing work of social movements. Victims of injustice are identified and their victim status is stressed to call attention to situations and circumstances that need to be addressed. Each of these elements can be identified within the establishment and development of the environmental justice frame in the US.

Capek (1993) provides one of the first analyses of the framing of environmental justice in the US, identifying its salient characteristics as it emerged from local community struggles over the siting and operation of toxic and waste sites in minority communities through the 1980s. She traces how local residents, mobilising against various perceived threats to their safety from pollution, leaks and contamination, began using a common language of environmental justice, giving them a political edge that was new and distinctive. Particularly important were, first, the way in which it tapped into the discourses of the civil rights movement, introducing issues of race and racism into environmental debate, and, second, the interplay between the scale of the specific local community struggles and the broader arguments and claims that emerged at a national level about the concentration of waste and toxic sites in minority communities. Benford (2005) in a later analysis argues that the initial discourse of environmental racism, which resonated with minority communities mobilising against risks to their safety and well-being, productively broadened to the environmental justice frame, which was more inclusive of the many forms of environmental discrimination that were being diagnosed. The language of environmental justice, he argues, was also more positively orientated, focusing on citizens' rights and on visions of what a more environmentally just world would be like. For Taylor (2000), the rapid growth in the use of the environmental justice frame in the US through the 1990s, as hundreds of groups formed around the country and a national movement emerged, even gave it the status of a 'master frame', a broad canvas that transcended the particularities of specific local disputes.

The movement achieved significant impacts in the 1990s both through local legal challenges to siting and other decisions, and through the lobbying of national policy-makers. Most significant in policy terms was concerted lobbying at federal level, which led to the creation of an Office of Environmental Justice within the Environmental Protection Agency (EPA) and the signing of Executive Order 12898 by President Clinton in 1994 requiring federal regulatory agencies to make environmental justice a part of all they do. In Box 2.1 the key provisions of the Executive Order and the story of its controversial and much criticised implementation within the EPA are outlined.

In the face of political change to a Republican administration in 2001 the environmental justice movement struggled to maintain its salience and momentum. A major backlash emerged from corporate interests, with political attacks being made by organisations and commentators working with very different frames and objectives (Benford 2005). 'Counterframes', deployed through what Faber (2008: 238) refers to as globalised tactics of the 'polluter-industrial complex', were for a while influential in criticising the economic impacts of environmental

Box 2.1 Executive Order 12898 and its troubled implementation

In 1994 President Clinton signed Executive Order 12898 as the culmination of a long period of campaigning and lobbying by environmental justice activists. The key requirement of this legislation is that 'each Federal agency shall make achieving environmental justice part of its mission by identifying and addressing, as appropriate, disproportionately high and adverse human health or environmental effects of its programs, policies, and activities on minority populations and low-income populations'. The passing of this legislation was seen as a major political achievement, but there has since been much debate about how the principles and requirements of the Order have been interpreted (or some would say reinterpreted), particularly within the Environmental Protection Agency (EPA). The EPA has been criticised from many directions, especially after 'interim guidance' was produced in 1998 on how it would assess whether pollution control permitting decisions made by state and local level agencies were in compliance with environmental justice policy (so-called Title IV decisions). Lyle (2000) provides a detailed account of the often vociferous reactions to this guidance. State and municipal-level interests objected to federal intervention in their local decision-making and the potential blocking of investment in job-creating industries. Industrial lobbies also weighed in with much force.

Having had its fingers burned by this furore, the EPA largely backed off from pushing environmental justice policy at a strategic level, leaving each region to develop its own approaches. Faber (2008) also argues that the EPA was deliberately denuded of resources during this period and undermined as a result of pressure from powerful corporate interests. An evaluation by the Office of the Inspector General in 2003 concluded in very strong terms that the EPA had neither fully implemented Order 12898 nor consistently integrated environmental justice into its day-to-day operations 'Although the Agency has been actively involved in implementing Executive Order 12898 for 10 years, it has not developed a clear vision or a comprehensive strategic plan, and has not established values, goals, expectations, and performance measurements' (Office of the Inspector General 2004: i). The consequence, they concluded, was an inconsistent approach by the EPA regional offices. Most fundamentally, the report argued that the EPA had reinterpreted the meaning and intention of the Order away from a focus on poor and minority populations towards seeking 'environmental justice for everyone'.

justice activism and undermining its progressive objectives. Some environmental justice groups, by then established in many communities across the US, faltered, but many survived through adapting their tactics and to some degree their profile of campaigns and activities.

With the election of President Obama in 2009, the environmental justice movement has found itself again more in step with the currents of political power, and there is some optimism about the potentially positive impacts of stimulus packages for deprived communities, and the greening of energy infrastructure (Holifield *et al.* 2010: 19).

Seven key characteristics of the US frame

There is much more that could be said about the development of environmental justice in the US, and elements of this story will figure in later chapters. However, from the account so far, and the body of analysis which underpins it, we can identify a set of seven key characteristics of the US environmental justice frame as it has evolved over the past 30 years:

1   It has emphasised a *politics of race*, reflecting its emergence from a history and infrastructure of grassroots civil rights activism. This made it not only an innovative and radical frame but also a new brand of environmentalism (Schlosberg 1999; Taylor 2000), involving a far more diverse constituency of activists than the traditional environmental movement, a diversity which has extended over time to include many different racial, ethnic and cultural groups. It would be wrong therefore to characterise the US environmental justice frame as being *only* about a politics of race (Faber 2008), as other forms of class and identity politics, including that of gender (Kurtz 2007; Stein 2004), have been involved, although the initial emphasis has arguably remained in place throughout.

2   It has maintained a focus on questions of *justice to people* in the environment (Agyeman *et al.* 2003: 327), rather than expressing a politicised concern for justice to nature – a separate question of 'ecological justice' in the categorisation of Low and Gleeson (1998). This anthropogenic placing of people and communities at the centre of the frame, particularly those who are marginalised economically and politically as well as environmentally, again distinguished environmental justice from the traditional framings of environmental groups in the US, which focused on wilderness and conservation concerns (Shrader-Frechette 2002).

3   In terms of the framing of its *environmental boundaries*, the early formulation of environmental justice was narrowly focused on forms of technological pollution, waste and risk – particularly those forms of 'environmental bad' associated with proposed new sitings of landfill, incinerators, chemical plants and the like. As we saw in Chapter 1, this narrowness has since given way to a far broader profile of environmental concerns, moving beyond environmental burdens to include access to environmental benefits

and resources of various forms (Mutz *et al.* 2002) and concerns which some argue could, or should, be classified as social rather than environmental (Benford 2005).

4   It has similarly evolved beyond an initial emphasis on issues of *distributive justice* to be more inclusive of other forms of normative claim and assertion. Environmental justice activism has always been concerned with more than distribution, including demands for participatory justice in particular (Schlosberg 2007; Shrader-Frechette 2002; Wenz 1988). Box 2.2, which reproduces the mission statement of the Little Village Environmental Justice Organization in Chicago, makes this clear with its strong emphasis on 'democracy in action', 'participation' and 'self-determination'. However, distributive claims – about who gets what in the environment – have dominated most representations of environmental justice in the US. This is partly because of the close association between the initial phases of activism and statistical studies which analysed patterns of distribution of environmental bads in relation to the racial and income profiles of affected communities, finding repeated patterns of bias and disproportionate concentration in poor, African-American and Hispanic areas (Bowen 2002; Brown 1995; Mohai and Saha 2006) and using the courts to challenge siting decisions which further reproduced these biases (see Chapter 3 for more specific discussion).

5   In diagnosing the causes of inequality and injustice, or assigning *blame and responsibility*, it has been focused on industry and corporate actors, and on the institutionalised (and racist) practices of the state. For example, both

---

Box 2.2 Mission statement of the Little Village Environmental Justice Organization

Our mission is to work with our families, co-workers, and neighbors to improve our environment and lives in Little Village and throughout Chicago through democracy in action. We work for a real voice in building democracy, including if, how, when and where any development of our communities takes place, as the basis for environmental, economic and social justice. Our environment is where we live, work, study, play and pray. We work with, not against, our Mother Earth and Nature to once again make our air healthy to breathe, our water safe to drink, and to free our earth from poisons to grow healthy foods.

We believe democracy means giving time and space for every voice to be heard and counted in everyday matters, full participation in all types of decision-making that affects our lives, and determining the future of our neighborhood and city.

We work to unite our community's talents, assets, and power to build a society that treats all of us equally: no matter what race, culture, ethnicity, age, or gender we are. In Unity we have the strength to forge economic, environmental and social justice to overcome the barriers of poverty that surround us and build self-determination.

We work for justice at home and abroad, connecting our local struggle for democracy with the global one and live by the principle that, as working and poor people of color, we have the right to control our lives and resources.

Source: http://lvejo.org/about/mission-statement.

Figure 2.2 Protest march by the Little Village Environmental Justice Organization in Chicago.
Source: Toban Black.

industry and government bodies have been blamed for inequitable siting decisions (see Chapter 4) and the operation of industrial installations to varying standards (Gouldson 2006).

6   It has been explicitly inclusive of multiple interconnected *scales of analysis*, but until recently these have been contained within the borders of the US. As noted earlier, a key strength of the environmental justice frame has been both the horizontal interconnections made between numerous local

grassroots struggles across the US and the vertical scaling up to national claims and regulatory settings (Kurtz 2002; Towers 2000). However, for some time these horizontal and vertical scalar connections remained bounded within national borders.

7   Whilst a broadly based environmental justice frame has been rooted in a vibrant social movement – as well as in the work of academics with whom the movement has been closely connected (Cable *et al.* 2005) – other versions of an environmental justice frame have emerged within the US *government and its agencies*, in part because of the success of activists in demanding policy attention (as outlined in Box 2.1). An essentially managerial framing has been adopted by the EPA and other agencies, one that is far more narrowly conceived than that of the activist community, and much criticised in its implementation (Block and Whitehead 1999; Faber 2008; Holifield 2001, 2004).

Whilst not exhaustive, these seven dimensions provide a sufficient characterisation with which we can proceed to examine how the use of an environmental justice frame has emerged in places outside of the US, across new networks and at a global scale.

The international travelling of the environmental justice frame

[I]f the environmental justice movement is to survive at all it must go global. It must go global, because the sources and causes of environmental inequality are global in their reach and impact.

(Brulle and Pellow 2005: 296)

The movement of the environmental justice frame beyond the borders of the US has happened over an extended period, although as Debbane and Keil (2004) show, through case studies based in Canada and South Africa, each particular case of transfer may happen relatively rapidly. The first manifestations can be found in the early to mid-1990s, with a more expansive diffusion taking place after 2000. A snapshot taken in 2010 provides an indication of how far the use of the language of environmental justice has reached. Table 2.1 lists the countries in which the specific term 'environmental justice' has been applied and written about in relation to indigenous environmental concerns, based on a search of academic and grey literature databases and web searches. This listing is indicative at best, as there are problems in relying on database and web searches; on the one hand, an environmental justice frame may be in use within a country without this having been written about (in English) or named precisely in this way, and, on the other, environmental justice may be used as a framework for academic analysis rather than being explicitly part of the discourse of those involved in activism or policy debates. It would also be wrong to interpret the adoption of an environmental justice framing as synonymous with the extension or

Table 2.1 Countries included in written material using an environmental justice frame

Region	Countries
Africa	Nigeria, Ghana, South Africa, Tanzania, Cameroon, Zambia, Angola, Mozambique
Asia	Taiwan, Israel, India, Singapore, Philippines
Australasia	Australia, New Zealand
Europe	United Kingdom, Germany, Ireland, Sweden, France, Spain, Belarus, Bulgaria, Hungary, Macedonia, Romania, Slovakia, Czech Republic, Latvia
North America	United States, Canada, Mexico
South and Central America	Brazil, Peru, Nicaragua, Ecuador, Columbia

development of an indigenous environmental justice *movement* – this being far more than a matter of framing. Even so, the list of 37 countries in Table 2.1 is extensive and demonstrates that the language of environmental justice (at least) has been in use in each of the major global regions and in some cases across many of the countries within these regions.

This indication of the scale and extent of environmental justice framing activity militates against simple generalisation or distillation of the mechanisms of diffusion and adoption that have been involved. It is clear though that deliberate transnational networking between environmental justice activist groups in different countries has been part of the story, paralleling wider trends across various forms of social movement (Routledge *et al.* 2006; Smith and Johnston 2002). For example, the Coalition for Environmental Justice, a civic action network of activists, lawyers and researchers from environmental and human rights organisations in Bulgaria, the Czech Republic, Hungary, Macedonia, Romania and Slovakia, was set up in 2003 to actively promote an environmental justice frame across Central and Eastern Europe. Network activities included linking up with environmental justice activists in the US to form a 'Transatlantic Initiative on Environmental Justice' in 2005 (Pellow *et al.* 2005), and the laying out of an agenda of key issues for Central and Eastern Europe, particularly focusing on the Roma who are discriminated against across the region (Steger 2007). There have been a number of such transnational initiatives using an environmental justice framing within other regions, such as South America (Carruthers 2008), or focused on particular environmental issues such as an 'anti-toxics' agenda (Pellow 2007).

Such networks have been significant in promoting diffusion from the US as well as generating interaction and learning between countries within regions – although these may not be the only mechanisms involved. A more in-depth analysis is required to understand not only how frames have emerged in new places, but also how an environmental justice framing once travelled becomes contextualised in its new cultural and political setting or becomes 'locally grounded' (Debbane and Keil 2004: 210). For this reason two cases will be examined in greater detail – the UK and South Africa; these cases contrast in many ways but, as we shall see, also show similarities in the contextualisation processes involved.

Environmental justice in the UK

In 1998 Dobson noted that, in comparison to the emphatic arrival of justice on the environmental agenda in the US, there had been no 'direct equivalent' in Britain (1998: 26). The closest contemporaneous parallel to the US experience had been the UK Black Environmental Network (BEN), which in the 1980s highlighted the white, middle-class nature of much environmentalism and worked with local black communities to develop environmental awareness and involvement in conservation work (Agyeman 1987). However, the BEN remained small scale and failed to mobilise any significant constituency of support or to develop a more radical campaigning profile. Similarly, whilst there was a history of opposition to the siting of 'toxic' and polluting facilities in the UK – including the formation of networking initiatives such as 'Community Lobby Opposing Unhealthy Tips' and 'Communities against Toxics' – these had failed to develop any form of collective agenda around justice arguments.

In contrast then to the grassroots emergence of environmental justice in the US, it was a mainstream and established environmental group, Friends of the Earth (FoE), that first started to work with an environmental justice frame in the UK. In the mid-1990s FoE had begun to develop a more socially aware and urban theme to its work (for example, related to fuel poverty issues) and to work in closer collaboration with social and development NGOs (for example, through the Real World Coalition formed in 1996), and an environmental justice framing fitted well with these developments. Through collaborations with academics working to formulate a UK environmental justice agenda (Stephens *et al.* 2001) and networking with US activists, there was both a drawing on the US environmental justice frame and a purposeful redefinition to fit the UK political context at the time. Bob Bullard, a self-described 'kick ass sociologist' and key activist in the US environmental justice movement, was brought to the UK as a guest speaker at a number of academic and NGO events. However, the agenda he laid out was very much reinterpreted in the UK situation. In particular, an opportunity was seen to make the environment more directly relevant to the recently installed 'New Labour' administration, which had campaigned strongly on social exclusion and inequality issues. A series of pamphlets and publications produced by NGOs, consultancies and political groups were highlighting the linkages between the New Labour government's priorities on social exclusion and the social dimensions of environmental concerns. Jacobs (1999), for example, in a pamphlet for the centre-left Fabian Society developed arguments around 'environmental exclusion' as a component of a new environmental modernisation agenda.

This combination of drawing on the US framing with the redefinition of its elements in the UK context can be seen across FoE's work at this time. Its first significant move was to undertake research which closely mirrored the US model of analysing the distribution of polluting industrial facilities to reveal biases in siting patterns (Friends of the Earth 2000, 2001). In making this step, it explicitly sought to convey a new style of gritty urban environmental concern (with some parallels to the positioning of activists in the US):

> this is the sharp end of social exclusion. On top of unemployment and crime these families and communities face the grime of industrial pollution. Here pollution is as far from a middle-class concern as it can get.
>
> (Friends of the Earth 2000: 2)

However, the research focused not on siting in relation to patterns of race or ethnicity, but on patterns of income – a social-class orientation which reflected the political context at the time and the lack of strong race-based civil rights mobilisation in the UK. Key agenda-setting publications developed by FoE in collaboration with academics and other NGOs are similarly positioned – introducing environmental justice by referring to the US experience, before then laying out a set of concerns that are quite distinct from the emphases of the US framing (Boardman *et al.* 1999; Stephens *et al.* 2001). These include international and intragenerational issues, inequalities in access to environmental resources including food, energy and water, transport needs and risks and aesthetic, mental and spiritual needs (such as quiet and access to the countryside). Again the lack of a distinct racial dimension is apparent in, for example, the way the foreword to one such publication positions the significant social divisions in class and age terms: 'environmental problems are serious and impact most heavily on the most vulnerable members of society, the old, the very young and the poor' (Boardman *et al.* 1999: 1).

Another distinctive feature of the diffusion of the environmental justice frame into the UK was its ready adoption into the discourses and policies of governmental bodies (Agyeman and Evans 2004; Bulkeley and Walker 2005). Whereas it took many years of concerted campaigning in the US to get the EPA to begin examining questions of environmental justice, its equivalent in the UK, the Environment Agency (EA), proactively did so early on as part of its own strategic political positioning (Chalmers and Colvin 2005). The EA included a debate on 'environmental equality' at its 2000 annual general meeting and initiated its own analysis of patterns of the social distribution of various environmental indicators in two commissioned research projects on 'environment and social justice' and 'addressing environmental inequalities'. These projects followed the classic US environmental justice method of statistically analysing spatial data sets at national and regional scales, but focused not on race but on social deprivation (for example, Walker *et al.* 2003, 2007).[1] The reframing work undertaken by the EA

---

1 I was involved in these projects and some of the results and methodological and political complexities involved are discussed in Chapter 3 in relation to river water quality, Chapter 6 on flooding and Chapter 7 on greenspace.

included not only its definition of relevant social and environmental concerns (including flooding and water quality, both central to its regulatory remit), but also the naming of the frame itself. Whilst clearly derived initially from the US environmental justice frame, the EA settled on naming its own agenda as being one of 'environmental inequalities'; this was seen as both less politically contentious and more aligned with familiar policy discourses such as that on 'health inequalities'. As outlined in Box 2.3, a position statement under this heading was produced in 2004 (Environment Agency 2004) that laid out some concerns and good intentions, but very little in terms of concrete commitments.

Box 2.3 Key aspects of the Environment Agency's position statement on environmental inequalities (Environment Agency for England and Wales 2004)

The EA position statement identifies two key issues – the variability of environmental quality between different areas and communities and the bias of the worst quality environments towards 'people who are socially and economically disadvantaged' and 'the most vulnerable and excluded in society'. This is seen as affecting health and well-being, adding to the burden of deprivation and limiting opportunities for people to improve their lives. The solutions called for position the EA as only one actor amongst many; 'government, business and society all have a role to play in addressing environmental inequalities at a national, regional and local level'. Five specific 'solutions' are called for:

1   A better understanding of environmental inequalities and the most effective ways of addressing them, through commissioning research.
2   Government policy that promotes a reduction in environmental inequalities through integrating environmental equality across all policies, evaluating new policies for their impacts on those living in the worst quality environment, and using tools such as equity assessments.
3   Addressing environmental inequalities through tackling disadvantage by building the environmental aspects of multiple deprivation into neighbourhood regeneration and health inequality programmes.
4   Regional and local planning that prevents environmental inequalities, through planning authorities carrying out Strategic Environmental Assessment, cumulative impact assessments, and addressing environmental inequalities in community plans.
5   Communities being supported and involved in decisions that affect their local environment, through information provision and involvement of people from deprived communities in decision-making.

A framing process going on more widely within government – led by the Department of Environment, Food and Rural Affairs which coordinated a cross-departmental working group on environment and social justice and commissioned a wide-ranging evidence review (Lucas *et al.* 2004) – also served to incorporate environmental justice ideas into pre-existing sustainable development framings (Agyeman and Evans 2004), rather than taking these up to form a distinctive new theme. Sustainable development was well established as a 'master frame' in the UK by the late 1990s and, through incorporating the interaction between the social and the environmental dimensions of sustainability, was seen to readily accommodate questions of social difference and inequality. In this vein the 1999 national sustainable development strategy stated that 'everyone should share the benefits of increased prosperity and a clean and safe environment ... Our needs must not be met by treating others, including future generations and people elsewhere in the world, unfairly' (UK Government 1999).

In various ways, then, through adoption and reframing environmental justice in the UK was contextualised into contemporary cultural and political conditions, 'moving from the margins to the mainstream' (Agyeman and Evans 2004: 159), but also arguably in the process being stripped of some of its more radical and distinctive qualities. Its adoption by 'elites' in existing established environmental groups and government agencies (Bulkeley and Walker 2005), its lack of grass-roots mobilisation and its renaming and incorporation into existing framings, each to some degree weakened the frame's substance and significance in comparison to the US version. This is brought home by the contrast between two self-named 'Environmental Justice Summits' held on either side of the Atlantic. The first, held in Washington in the US in 1991, brought together over 650 representatives of grassroots organisations from around the country working within an environmental justice frame; the second, held in London in 2008 and organised by Capacity Global (the only clear example of a group organised around an environmental justice framing in the UK) with funding support from a government department, involved 50 people, most of whom were academics and representatives of government agencies or of national-level NGOs and consultancies.

Whilst this analysis may characterise the London-focused picture in the UK, in Scotland things have been a little different, demonstrating that forms of contextualisation can take place at levels below that of the state. In Scotland political opportunities were presented by the devolution of substantial responsibilities of governance to the Scottish Parliament in 1999. Friends of the Earth Scotland (FoES) deliberately chose this moment to adopt a more substantial and radical environmental justice campaign than elsewhere in the UK (Scandrett 2007) which interlinked local and global issues, supporting this with various forms of training and networking activity intended to empower local-level activism (Dunion 2003). Having been promoted strongly by FoES, a version of the environmental justice frame, focused in this case on local environmental conditions (or 'environmental incivilities'), also moved into government, with Jack McConnell, Scotland's first minister, declaring in 2002: 'I am clear that the gap between the haves and have-nots is not just an economic issue. For quality of life,

Figure 2.3 Friends of the Earth Scotland environmental justice campaign leaflet, 2001.

closing the gap demands environmental justice too. That is why I said ... that environment and social justice would be the themes driving our policies and priorities.' (McConnell 2002). This speech, briefly at least, catalysed attention and promoted the explicit use of the term environmental justice as a policy objective, with a dedicated team being established in the Scottish Executive. Various resource commitments were made to fund research (Curtice *et al.* 2005; Fairburn *et al.* 2005), support community action and review the implications for planning legislation and pollution regulation (Jackson and Illsley 2007; Poustie 2004). The 2002 Scottish sustainable development strategy was also explicit in its appeals to environmental justice, stating that 'sustainable development is about combining economic progress with social and environmental justice ... we should have regard for others who do not have access to the same level of resources, and the wealth generated' (Scottish Executive 2002). Despite such rhetorical commitments, Scandrett (2007) is critical of the way in which the environmental justice frame in Scottish policy has evolved, in particular its failure to in any way challenge the interests of capital. The election of the Scottish National Party to power in 2007 led to new policy discourses around environmental concerns as the new administration sought to distinguish itself from the old.

Consequently, the environmental justice frame has not been sustained in policy and the campaigning work of FoES has also become much less strongly orientated around this.

Environmental justice in South Africa

The first traces of the emergence of an environmental justice frame in South Africa can be found in 1992–3, a few years before those in the UK. Key events referred to in various accounts include an Earthlife International Conference in 1992, leading to the formation of the Environmental Justice Networking Forum (EJNF) in 1994, which has since grown into a network with over 400 members across a diversity of civil society organisations (Duma 2007). Here the US influence appears to have been significant in various ways. A participant account by Kalan and Peek (2005) traces early initiatives by students from South Africa studying in the US to connect the environmental struggles in the US with those of their home country. The 'South African Exchange Programme on Environmental Justice' (SAEPEJ) sought to develop two-way exchanges of various kinds – exchanges of information and research, the meeting of people from grassroots organisations and communities mobilising around similar environmental problems in the US and South Africa, and even the collection of samples of toxins from South Africa that were then taken for analysis in labs in the US. For Bobby Peek, who formed groundWork in 1999 as a group seeking to promote environmental justice activism both within South Africa and more broadly across the region, the link to the US was crucial: 'the language that was appearing in the civil rights movement and around the environmental justice movement during the late 1970s and early 1980s was something that came to South Africa in the late 1980s and early 1990s' (ibid.: 261). The two-way nature of the exchange is also clear, however, with learning about organising at a local level in South Africa being instructive for US activists, and the understanding that 'these things happen globally' (ibid.: 260) pushing US groups towards a more international perspective. Their account shows how the movement of knowledge and commitments embodied in particular people can be important in frame diffusion (Faber 2005).

Struggles against the operation of oil refineries and other sources of pollution in the heavily industrialised basin of South Durban were also significant in giving a focus and profile to the emergency of environmental justice activism. As Barnett and Scott (2007) trace in some detail, the South Durban concentration of industrial development took shape during the apartheid era, with non-white communities being forcibly relocated into the area in the 1950s and 1960s under the Group Areas Act. Concerns about high levels of ground, air and water pollution were already long-standing, and there was a strong profile of local civic organisation and political activism which had inputs into the ANC's environmental policy in the early 1990s. This provided the foundation for the formation of the umbrella organisation the South Durban Community Environmental Alliance (SDCEA) in 1996, which took up an agenda explicitly using the language of environmental justice that was by then circulating within NGO networks.

Over subsequent years the work of the SDCEA has made 'South Durban's two oil refineries (two of only four in the country) emblematic of environmental justice conflict' (ibid.: 2616) and a model for how to engage in community mobilisation in other pollution 'hotspots' around the country.

In the environmental justice frame that has emerged in South Africa there are many parallels with the US (Debbane and Keil 2004; McDonald 2002). Most significantly, the connections between the civil rights movement in the US and anti-apartheid struggles in South Africa meant that the discourse of environmental racism resonated strongly in a country where the racialisation of space had been institutionally organised and maintained through state power. Other parallels included the focus on toxic and polluting activities and on anti-corporate campaigns, and the deliberate contrast drawn between new activist discourses and traditional South African environmental concerns of wilderness and nature conservation based in colonial and post-colonial ideology (Martinez-Allier 2002).

The post-apartheid arrival of democracy in South Africa in 1994 had the task of addressing deep inequalities, including environmental inequalities of various forms which discriminated against the majority black population. The Bill of Rights of the South African Constitution accordingly included several statements of environmental rights: 'everyone has the right to have access to sufficient food and water ... an environment that is not harmful to their health or well-being ... to have the environment protected, for the benefit of present and future generations' (Republic of South Africa 1996: s.27.1, s.24).

Whilst this positioning of environmental rights at the heart of the new constitution appeared a powerful assertion of the environmental justice frame, a number of observers have critiqued the way that environmental management has since been practised. They have particularly pointed out the lack of procedural as well as distributive justice – in the form of meaningful opportunities to participate in decision-making – and the obstacles presented by other more powerful framings of environmental governance. As in the UK, Patel (2006) argues that the sustainability frame, which became rapidly established in post-apartheid South Africa (O'Riordan *et al.* 2000), has been dominant, often interpreted in technical and managerial ways that have failed to shake off the legacies of established colonial approaches to conservation and environmental management. She contends that, consequently, social and environmental justice dimensions have failed to be addressed within sustainability programmes and that the use of standard environmental assessment tools has failed to consider distributional consequences (Patel 2009). Bond (2000) similarly sees a neoliberal 'ecological modernization' perspective at work, overriding the individual rights supposedly protected by the constitution, whilst Oelofse *et al.* (2006) point to both a reliance on technocentric scientific approaches and an institutional implementation deficit as limiting the way that environmental objectives have been pursued. Debbane and Keil (2004) point to particular tensions of these forms in the case of management of water and water supply in the post-apartheid period. In these ways the enshrining of environmental justice rights in the constitution has not, as yet at least, had a significant impact on established dominant policy framings of key actors, such as the Department for Environment and Tourism, or on the deeply embedded structural legacies of apartheid (Kalan and Peek 2005). Similar tensions are identified in research using an environmental justice frame focused on land reform in conservation areas (Geisler and Letsoalo 2000) and urban resettlement programmes (Dixon and Ramutsindela 2006).

Amongst environmental justice activists, however, the focus on questions of inequality and justice and the distributive and procedural rights of historically marginalised township communities has remained in place, but the scope of their framings has extended into a diversity of socio-environmental issues affecting the lives of ordinary people (McDonald 2002). These have included the provision of basic resource and infrastructural needs – such as water and electricity (Bond 2000; Debbane and Keil 2004), health and safety for workers in the mining sector, and the health risks of asbestos and herbicides (Martinez-Allier 2002). For example, the Environmental Justice Networking Forum has approximately 400 members across a diversity of civil society organizations, including faith-based, trade union, women's, youth and children's organizations, and a stated profile of concerns that include mining, food security, energy, waste, water and biodiversity – although it has struggled, as have many such initiatives, to maintain its infrastructure and resource base (Duma 2007). These groups have also readily worked with both an environmental justice and a sustainability framing, strategically shifting the labelling they use in different contexts and interactions.

In part as a consequence, McDonald (2005) argues that there have been significant differences of opinion amongst environmental justice groups in South Africa over the importance of race, gender and class as social framings and the potential to achieve meaningful reform within a market-based economy. Barnett and Scott (2007), in their analysis specifically of the work of the South Durban Community Environmental Alliance, identify many such tensions, for example, in the potential for the group to become co-opted through inclusion in formulaic decision-making processes and in its relationship with international donor NGOs pushing for cooperative rather than confrontational ways of working with the state and business. In moving towards partnership working and procedural inclusion, the SDCEA has faced major challenges in reconciling these strategies with foundational demands for historical redress and accountability for discrimination and environmental harm experienced over the long history of apartheid rule.

Comparisons and contextualisation

In the examples of the UK and South Africa we can see various forms of contextualisation, or grounding of the environmental justice frame. There are similarities and contrasts between the experiences in each country. Similarities include:

- a clear reference to and learning from the environmental justice frame in the US, and examples of international networking and interactions which promoted frame diffusion;

- an environmental justice frame having been taken up not only as a collective action frame within social movements, but also by government bodies who have introduced their own meanings and interpretations;
- differences having opened up between activist and governmental framings in terms of their constituent elements, specific languages and application;
- environmental justice being set alongside or within an existing sustainable development master frame, both by campaign groups and governments, leading to tensions as to their compatibility and relative importance;
- the importance of particular political events – new democratic institutions, devolution, new administrations – in providing openings for the introduction of an environmental justice frame into political debate and policy commitments.

Contrasts in the contextualisation of environmental justice in the two countries centre particularly on the extent to which the frame has been part of grassroots networking and has encompassed a discourse of environmental racism. Faber (2005) identifies a number of different competing discourses dominating environmental justice politics, only one of which is based around racial identity. In the UK the environmental justice frame has been promoted primarily by a mainstream environmental NGO. A wide range of socio-environmental issues have been included within the frame but without an emphasis on racial or ethnic identity politics. In Faber's (2005) categorisation, a 'socialist politics' has dominated, focused primarily on shared material interests or social class – although in Scotland there has been more substantial grassroots activism and the politics have had something of a nationalist flavour. In South Africa, there is more evidence of environmental justice emerging as a frame for grassroots mobilisations, following more closely the US trajectory, and including race as a key, if not dominant, discourse. Because of this, and its foundation in anti-apartheid politics, the environmental justice frame in South Africa has maintained a more radical edge, with activists positioned more clearly in opposition to rather than in consensus with governmental actors, although strategic tensions around this have been identified.

This profile of similarities and differences both between the two cases, and in comparison with the US, is sufficient to demonstrate that the environmental justice frame is not singular but flexible and dynamic, open to reconstruction as it moves both in space and time. In this way Williams and Mawdsley (2006) argue that the geography of environmental justice matters; it has to be defined within the context of each site in which it is used rather than being readily universalised under only one conceptualisation. As environmental justice globalises, its initial meaning derived from the US context is not simply reproduced, although neither is it entirely abandoned. There are a growing number of national contexts now being discussed within the literature which demonstrate this. For example, there is the way in which environmental justice in Israel engages with the dominance of security concerns and the intensely politicised status of the Arab minority population (Shmueli 2008), in Brazil with post-colonial legacies

(Souza 2008), and in Taiwan with the country's emergence from a military dictatorship and a growing movement for indigenous rights (Huang and Hwang 2009).

Before returning to discuss further the implications of these observations, we can now move to examine how the environmental justice frame has globalised not only 'horizontally' in space but also 'vertically' in its scales of concern.

Environmental justice framings of global issues

> The environmental justice movement is potentially of great importance, provided it learns to speak not only for the minorities inside the USA but also for the majorities outside the USA (which locally are not always defined racially) and provided it gets involved in issues such as biopiracy and biosafety, or climate change, beyond local instances of pollution.
>
> (Martinez-Allier 2002: 14)

For those looking from the outside, a striking feature of the US environmental justice frame in its earlier manifestations was its introspection. As already noted, its dominant concerns were with 'who got what' within the cities and regions of the US (Dobson 1988), not with questions of distribution, disproportionate impact or marginalisation extending beyond the borders of the US to encompass people elsewhere and the implications of international or global environmental processes (Newell 2005). For some observers, enthusiastic in other ways about the new form and constituency of environmentalism that had emerged in the US, this was a significant limitation (Martinez-Allier 2002), as it was failing to grapple with the justice issues which were paramount for many environmental and social advocates outside of the US and already situated within a sustainable development framing.

The shift towards environmental justice framing beginning to vertically 'upscale' its scope of concerns is not disconnected from the horizontal travelling of ideas and meanings discussed in the previous section. Part of the contextualisation processes that take place in frame movement involves redefinition of the scope and reach of the frame, and this redefinition can readily encompass not just indigenous local and national issues but also international and global ones. For example, in Scotland, when Friends of the Earth first formulated its environmental justice campaign theme, it adopted a definition of environmental justice which neatly and succinctly expressed the simultaneous local and global reach of justice issues 'no less than a decent environment for all: no more than a fair share of the Earth's resources' (Friends of the Earth Scotland 1999). Here justice is conceived in terms of both local rights to environmental quality and importantly also global responsibilities deriving from patterns of consumption (Dunion and Scandrett 2003). Similarly, when environmental justice became the framing for transnational activist networks, this was not restricted in substantive terms to connecting up mobilisations focused on local disputes over facility siting, access to clean water and so on. Transnational networks also positioned responsibilities for harm in distant internationally dispersed locations firmly within the frame, connecting globalised economic and political relations with their environmental consequences (Pellow 2006). For example, the agenda of the Coalition for Environmental Justice transnational network in Central and Eastern Europe includes the exporting of risks from richer to poorer countries alongside a range of country-specific concerns (Steger 2007).

Looking, then, across the international and global scope of the various environmental justice framings that have been adopted internationally – as well as the organic development of the frame in the US, which in the late 1990s increasingly began to look beyond its own borders (Bullard 2005; Pellow 2006) – a diversity of internationally structured issues can be identified. Two of these will be discussed in later chapters – the international movement and disposal of hazardous and electronic wastes in Chapter 4, and global issues of climate change and climate justice in Chapter 8. Trade agreements are an important third example. The engagement of environmental justice activism with international trade and trade policy has been most directly analysed by Newell (2007) in the context of various forms of mobilisation in South America against both continent-wide and sub-regional trade agreements. He argues that groups working with, or drawing in part on, an environmental justice framing have been able to mount a stronger environmental critique of regional trade integration in the Americas which is far more grounded in justice to people and communities than the nature conservation agendas advanced by mainstream environmentalists involved in trade agreement campaigning. This deliberately atypical form of environmentalism, grounded in 'campesino' and indigenous peoples' movements (and thereby claiming a broad constituency of support), has been driven by the local experience of living with neoliberal approaches to the control of resource rights and basic services such as water provision. Furthermore, he argues that an environmental justice frame has provided the basis for critiquing the procedural elements of trade policy, 'who participates, on whose behalf and who gains from trade policy and at whose expense' (ibid.: 238). Even where trade agreements have in principle conceded greater transparency and been opened up to a greater diversity of voices, the practices of involvement have been shown to be exclusionary and inaccessible to groups with a weaker resource base. Schlosberg (2007) makes a similar point in arguing that groups mobilised against global trade agreements in various parts of the world have been concerned not only with inequalities in the distribution of consequent environmental bads (pollution, waste and resource depletion), but also with matters of social and cultural recognition and participatory justice (see next chapter).

This and other examples of international-scale environmental justice concern show again how the frame has been open to evolution and recontextualisation over time. In the process of 'scaling up', other dimensions have also had to evolve, further distancing these evolved framings from the characteristics of the early US collective action frame. As we shall see in Chapter 8, with climate change in particular the assignment of blame and responsibility has extended beyond corporate and state actors to include the consumption practices of nations and their citizens, a crucial development for more directly revealing the structural fault lines in relationships between the Global North and South. Climate change has also extended the driving concern for justice to people to include those who form part of future as well as current generations (Schlosberg 2007). The established prognostic demands for action by national and local levels of government have also had to extend further, to include demands for action on environmental justice by transnational intergovernmental regimes.

The implications of 'going global'

So it is clear that environmental justice mobilisation and framing has now taken on an increasingly global form and perspective and that its reach extends far beyond the US and hence into very different socio-political circumstances. Both Joan Martinez-Allier (from Spain) and David Pellow (from the US), whose appeals for a more global environmental justice have been quoted earlier in this chapter, would approve. It is becoming an international master frame that, as Dawson (2000) argues, does not appear to require a particular political or economic context in which to flourish. In moving horizontally across space, vertically across scales and temporally as socio-environmental and political conditions have shifted, the environmental justice frame has shown the capacity to take on alternative emphases and to evolve and re-contextualise. Sometimes in both its horizontal and its vertical movement the environmental justice frame is proving instrumental in identifying new concerns and new material cases of inequality and injustice. It is more often though becoming attached to existing local, regional and international issues, framing and labelling these as matters of justice and thereby identifying them as part of wider systemic processes and wider demands for fairness and the protection of basic needs and rights (Schroeder *et al.* 2008).

In some ways it is ironic that environmental justice framing has emerged from the US, a country so deeply implicated in patterns of economic and environmental exploitation around the world, and in the causes of global scale problems such as climate change. Indeed this has itself created some difficulties and tensions for activists in countries such as South Africa that have strategically not wanted to be seen to be simply following a US-created discourse and model of campaigning (Kalan and Peek 2005). However, it has become clear from the preceding analysis that whilst the early US experience and networking with US activists have been influential, there is a large degree of local reinterpretation and reframing going on. As Debbane and Keil (2004) argue, and as discussed in the previous chapter, this demands a relative and scaled understanding of what constitutes environmental justice rather than one based on notions of universality and conformity. That is not to say that the environmental justice frame is born anew in each place it emerges, or that it has evolved out of all recognition from where it began. There are clear common reference points – around, for example, the incorporation of core demands for distributive and procedural justice, and, Schlosberg (2007) has argued, for individual and community-level recognition and capability to function (see Chapter 3) – but the ways in which these are interpreted, combined and operationalised are open to variety and diversity. In a similar vein, Schroeder *et al.* (2008) argue that the core issues at the heart of environmental justice struggles, wherever they are found, are universal, part of broader patterns of distributive, procedural and racial injustice with global significance.

Whilst the capacity to co-evolve with socio-environmental and political change and to go global can be seen as both positive and necessary for environmental justice framings to continue to be relevant and to 'do work' for activists groups and the communities they represent, the analysis in this chapter has also identified tensions within this process. There is demonstrable scope for the radical edge of claims for environmental justice and the realisation of environmental rights to become blunted through reframing, relabelling and incorporation into the managerialist frameworks of government bodies. The case of the UK is instructive here. Much has been learnt, the evidence base on environment and social difference has progressed considerably (as we shall see in later chapters), and there is yet scope for a more strident justice discourse to emerge – particularly around international and global issues. It would clearly be wrong, though, to talk about an environmental justice *movement* in the UK, or indeed anywhere else in Western Europe, where managerialist and technocratic versions have tended to predominate.

The interaction between environmental justice and sustainable development framings is also interesting territory. For some observers and activists, their coming together is absolutely necessary and productive under the framing of 'just sustainability' (Agyeman and Evans 2003), but, as we have seen in South Africa, for others the tendency of sustainability perspectives to emphasise compatibility with the market, consensus approaches and ecological modernisation solutions can mean that questions of inequality and impacts on vulnerable and excluded groups are too easily downplayed, if not pushed aside.

Summary

Framing has been used in this chapter as a concept to examine how the environmental justice frame first emerged in a US context and how this gave it a series of distinctive characteristics that shaped its scope, priorities and assumptions. We have then traced how the use of the term has since globalised in two dimensions – horizontally in emerging in other places and cultures around the world, and vertically in becoming concerned with international and global justice issues. This, it has been argued, has implications for how we understand environmental justice as a frame that has multiple forms and is contextualised within the settings of its particular uses. Whilst the focus in this chapter has been predominantly on environmental justice as a frame for political activism – or collective action – we have seen how environmental justice discourse has also been translated into policy domains, with consequences for how it is then interpreted and understood.

Further reading

There are a number of edited books that have brought together contributions from different parts of the world and emphasised the international dimensions of environmental justice – these include Agyeman *et al.* (2003), Bullard (2005), Holifield *et al.* (2010) and an excellent recent addition: Carmin and Agyeman (2011). Several journal special issues have also followed this approach – *Geoforum* (2006, vol. 37, no. 5); *Society and Natural Resources* (2008, vol. 21, no. 7); *Antipode* (2009, vol. 41, no. 4). The work of Martinez-Allier (2002) on the 'environmentalism of the poor' has a broad international scope. The journal *Environmental Justice* (www.liebertpub.com) also has an international scope and has included short papers discussing environmental justice in various countries, including Taiwan, Brazil and Israel.

For material specifically on the UK context, see a special issue of the journal *Local Environment* (2005, vol. 10, no. 4) and reports and presentations available at the research and resource site at Lancaster University: http://geography.lancs.ac.uk/EnvJustice/. Capacity Global, a UK-based environmental justice group, has a useful website at www.capacity.org.uk/.

Environmental justice activism in South Africa is examined in a book by McDonald (2002), as well as various journal papers, notably Leonard and Pelling (2010) and Barnett and Scott (2007).

3   Making claims
Justice, evidence and process

Many dimensions of the intertwining of environment and social difference and their enrolment within an environmental justice frame have already been introduced. In this chapter I focus directly on this multi-dimensionality and explore it through a simple framework that identifies the different elements involved when researchers, activists, policy-makers and others make claims about environmental justice concerns. Claim-making was identified in Chapter 1 as one of the themes running through the book and as central to the analytical perspective I have taken. The aim of this chapter is to examine the constituent elements of environmental justice claim-making and how they interrelate, and through this to consider key concepts and ideas that span different disciplinary perspectives. This chapter is more theoretically orientated and therefore distanced from the materiality of environment and social difference than others. Subsequent chapters provide far more real-world substance and will illustrate and explicate much of what is introduced here.

The framework has three elements – justice, evidence and process – and there is some challenge in attempting to cover each of these in one framework, and in one book chapter. Each of the three elements, and their interrelation, need to be encompassed if we are to have a full suite of resources for the critical analysis of instances of claim-making, revealing the explicit and implicit mean-ings, values and ideologies that are at work. There are few academic accounts of environmental justice that do this. There are many excellent treatments of justice theories, principles and concepts in the literature (on justice in general, as well as more specifically on environmental justice) and this work will be drawn on in what follows. Less widely available have been accounts which take an analytical approach to understanding and categorising the forms and dimensions of evidence that are brought to bear in claim-making and how these interplay with different notions of justice – although Liu (2001), Foreman (1998) and Shrader-Frechette (2002) all go some way in this direction. Similarly, frameworks for understanding the processes through which inequalities and injustice are produced have tended to be divorced from a concern for critically evaluating evidence, and to some degree from the articulation of particular notions of justice and injustice.

I will begin with an outline introduction to the three elements of claim-making, how, why and to what degree they are distinct, and how they are interrelated. Justice, evidence and process will then be examined in turn.

The three elements of claim-making

The framework around which this chapter is organised has three interrelated elements. The main axis, as presented in Figure 3.1, makes a distinction and link-age between two types of claims – claims about evidence, or *how things are*, and claims about justice, or *how things ought to be*. This 'is–ought' distinction was explained in Chapter 1 as being useful (to me and others) for distinguishing between *inequality* as a descriptive term and *justice* as a normative term. Another formulation of this duality is between *evidence*, on the one hand, and *reasoning*, on the other; when evidence of inequality is tied to reasoning about why that inequality is wrong and fails to satisfy one or more justice principle, then the central formulation of a claim of injustice is in place. Connected to this main axis is the third element: claims about process, or *why things are how they are*. These claims are explanatory, seeking to identify the processes through which inequalities and injustices are produced and reproduced.

Three preliminary points about the interconnection of these three elements need to be made. First, it is not necessarily the case that all three will be, or have to be, in place when environmental justice claims are being made. For example, there are many examples of environmental justice research papers which focus on providing evidence, often of patterns of distributive inequality, without then moving to a reasoned case as to whether or not there is an injustice involved. This is still claim-making but of a restricted form. Instances can also be observed of strident claims of injustice which do not then provide much in the way of evidence to back them up, and of accounts of inequality which do not provide an explanation of underlying processes of production and causation. However, when justice claims are part of political activity, then each of the three elements and, crucially, their strategic interplay *will* often be in action. The framing work of

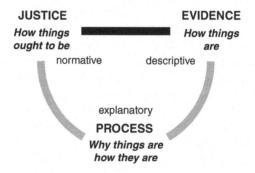

Figure 3.1 The three elements of environmental justice claim-making.

environmental justice groups discussed in the last chapter readily encompasses each of these elements, both in marking out a general way of understanding socio-environmental concerns and in campaigning about specific cases and disputes.

Second, there is similarly no presumed sequencing or flow between the three elements in Figure 3.1 – there are deliberately no arrows on the interconnecting lines. It is not necessarily the case, for example, that evidence will always come first, to be followed by the determination of whether or not and why injustice exists. This is far too dry and methodical an assumption. Instead, in practice, we might well see an initial expectation of injustice – based, for example, on viewing discrimination as a fundamental, structuring process – which is then followed by a seeking out of evidence with which to substantiate this expectation. Or the converse, a reluctance to accept that there can be any injustice, which is then rationalised by a search for evidence to show that there is no inequality to be addressed. Or indeed there may be cycling and circling between the different elements as claim-making develops and evolves within a particular context.

Third, it is crucial to guard against misinterpreting the distinction made between evidential description and normative reasoning in this framework. This must not be taken to imply a separation between fact and value, that values are involved in reasoning about justice, neatly demarcated from the descriptive 'facts of the case'. Such a separation is not intended or tenable. As ably demonstrated across a diversity of literature, 'evidence' is not something given, a set of truths waiting simply to be revealed. Rather, it is constructed and produced in ways that reflect the routines, epistemological predispositions, cultures and values of those doing the producing. Such considerations particularly apply to seeing science and scientific approaches as social constructions, based on certain shared assumptions, routines and practices which are represented as embodying rigour and rationality, but socially and culturally constructed none the less (Eden 1999; Gieryn 1999; Jamison 2001). Following this line of argument correctly positions evidence as a claim for knowledge, and therefore very much part of the process and politics of claim-making. As noted in Chapter 1, but worth repetition, the adoption of a constructivist and critical stance about knowledge, and scientific knowledge in particular, does not then mean that all becomes relational. As Forsyth (2003: 11) argues, 'it is quite possible to criticize many statements made by science while still believing in environmental realism (or the existence of a real world out there)', and clearly a concern for environment and social difference necessarily implies a belief that such things do exist and that their combinations can be harmful to human well-being. But being clear about how that knowledge is socially produced promotes a healthy scepticism about evidence and proof, and foregrounds a necessary politics about what we choose to give credit to and respect and what we choose to downplay or reject.

Having dealt with these preliminary but important qualifications, we can now engage with the complexities that hide within the preliminary simple formulation of the framework, starting first with justice concepts.

Justice concepts: how things ought to be

The argument that there is a multiplicity of justice concepts relevant to the theory and practice of environmental justice was made in Chapter 1. As environmental justice has evolved and moved around the globe, through the processes discussed in Chapter 2, it has become especially clear that a plural, multivalent understanding of the normative reasoning that goes on around the relation between environment and social difference is needed. These arguments are not reiterated here, but they underpin the structuring of this section, which draws substantially on Schlosberg (2004, 2007) and the justice theorists that he himself draws on. Only an outline is provided and readers are encouraged to go to the original sources, as well as to Wenz (1988), Dobson (1988), Shrader-Frechette (2002) and Bell (2004), for more thorough and substantial accounts.

Distributive justice

One of the few consistent features across the various definitions of environmental justice reviewed in Chapter 1 (see Table 1.1) was a concern for distributive justice – or who gets what in the environment. Distributive justice has been characterised as the 'chief topic' of environmental justice concerns (Wenz 1988: 4), the 'substantive justice' that matters in a material sense in terms of the burdens and benefits that are received (Bell 2004), and the 'fundamental question' of all justice theory (Brighouse 2004). More subtly, Schlosberg (2004: 529) argues that distributive justice is always central, but that our understanding of what constitutes and crucially produces injustice (linking to matters of process in Figure 3.1) is more complete and satisfactory when other concepts of justice are also brought to bear (see later discussion).

Bell (2004) lays out three questions that need to be addressed in order to construct a claim about distributive justice:

1   **Who are the recipients of environmental justice?** This involves determining a 'community of justice' or who matters when we think about the distribution of environmental benefits and burdens. This definition has important spatial and temporal dimensions. A 'community of justice' can be defined in narrow terms as, for example, the current population of a nation (thereby spatially intranational and temporally intragenerational), which was the formulation of the early phases of the US environmental justice movement (Dobson 1988). Or it can be defined in a more expansive and cosmopolitan manner (Caney 2007) to include populations in other nations and future generations (thereby international and intergenerational), which, as explored in the last chapter, has become more common as the environmental justice frame has evolved. For climate change in particular (see Chapter 8) how the community of justice is defined is crucial to the formulation of what justice in climate mitigation and adaptation becomes – and is therefore inevitably itself a source of much contention. Using political boundaries for spatial

Table 3.1 Issues in defining the appropriate metric of distributive justice for each of the topics of later chapters

Topics	Issues in defining what is to be distributed
Waste	What type of wastes or waste facilities? Is it proximity to waste facilities that matters and, if so, how is this to be measured? How about patterns of waste production?
Air pollution	Is it the diminishing of air quality or emissions of air pollutants that is to be distributed? And by what measure – ambient air quality, specific source emissions or levels of personal exposure; average levels, peak levels or exceedences of standards?
Flooding	Is it a level of exposure to potential flooding that is to be distributed, likelihood of being flooded or flood impacts? How about the distribution of investment in flood defences or in preparedness capacity?
Greenspace	How is greenspace to be defined given that it can take many different forms? What qualities are to be accounted for? What use values of greenspace are deemed to be important – visual impact, exercise and play, relaxation?
Climate change	What measures of carbon emissions are to be used – absolute levels, per capita levels – and at what scales? How about historic emissions? How might the distribution of climate change impacts be captured and at what scales?

    definition can be problematic where, for example, indigenous groups have historic patterns of settlement that cross national borders – see Vermeylen and Walker (2011).

2   **What is to be distributed?** We have already seen in Chapter 1 that contemporary understandings of environmental justice encompass a wide diversity of environmental features, including both burdens (air pollution, flood risk, noise, waste) and benefits (access to water, greenspace, energy consumption and services) – although the distinction between benefits and burdens is malleable (clean air is a benefit, air pollution a burden). Particular features or materials may also shift from benefit to burden depending on their context and evaluation. For example, energy consumption can be viewed as simultaneously a benefit in providing essential energy services and a burden in contributing to carbon emissions, and its distribution can be at issue in both respects (Boardman *et al.* 1999). Flooding can similarly be a positive resource for some and a threat to others (see Chapter 6). Schroeder *et al.* (2008) argue, in the context of environmental justice cases in the developing world, that the concepts of benefits and burdens 'are always relative, both in absolute terms and with respect to any particular group of potential resource users' (p. 550). There are also crucial subtleties in defining exactly what is to be distributed. As outlined in Table 3.1, for each of the topics of the following chapters there are issues in resolving what is

to be distributed. In each case different forms of complexity and optionality are revealed.

The question of what is to be distributed also clearly relates to the evidence that would then be needed to make evaluations and judgements. The ideal metric of distribution may not be available in existing data sets, and there can be many situations in which the primary distributive concern is not itself directly measurable, so some form of proxy indicator of distribution has to be used instead (for example, proximity to a source of pollution as an indicator of pollution impacts on health and well-being). Issues involved in relying on proxies are examined more closely later.

3   **What is the principle of distribution?** There are many options available here. Justice scholarship has debated these at length, and different positions on what constitutes the best or right principle of justice have provided the basis for deep political and ethical disagreement. Looking specifically to the claims of environmental justice advocates, Bell (2004) identifies three principles as generally being applied: first, a 'principle of equality', which might mean, for example, the equal distribution of waste sites across a terri-tory, or the equal per capita distribution of carbon consumption; second, a 'principle of equality plus a guaranteed standard', where inequality needs to be removed but also a standard of environmental equality ensured for all (such as a basic standard of air quality or right to clean water); third, 'a guar-anteed minimum with variation above that minimum according to personal income and spending choices', in which beyond an ensured minimum people can reasonably express their preferences in different ways. He uses the example of rights of access to the countryside or of living in a good quality environment to focus on how the failure to ensure minimum standards is the issue, rather than inequality per se. Bell also argues that a Rawlsian concep-tion of 'justice as fairness' can be readily extended to include environmental justice concerns.

These different principles are each theoretically applicable and relevant, but we need to do more to produce a sufficient account of the range of ways in which distributive justice can be interpreted within environmental justice claim-making. A pluralistic view of justice finds necessary diversity in the criteria of distribution to be applied, as 'there are many different social goods (and evils) whose distribu-tion is a matter of justice, with each kind of good having its own particular crite-rion of distribution' (Miller 1995: 2), an argument readily applied to the wide array of environmental goods and bads that we have already seen are at issue. Furthermore, there is much power in Walzer's (1983) contention that naming and giving meaning to any particular good or bad is a social process, therefore particular rather than universal. The criterion of just distribution should for this reason be expected to reflect the different meanings of goods and bads that emerge in particular contexts and, in part for this reason, should also be expected to be contested. Urban greenspace, as we shall see in Chapter 7, is not always and everywhere seen as an equally good thing, or valued for the same reasons.

What for some is an exciting place to play and 'hang out' can be a threatening, hostile and unsafe environment for others. We should not therefore expect the grounds for claiming the injustice of an unequal distribution of access to green-space to be agreed or consistently applied. Taking another example, Hillman (2004) ably demonstrates diversity in environmental justice concepts by identifying six alternative 'underlying justice principles' at work in priority setting for stream rehabilitation in Australia – see Box 3.1.

Similar listings of alternative distributive justice principles have been generated for other concerns, including climate change and carbon reduction

---

Box 3.1 Alternative principles of justice in priority setting for stream rehabilitation

Mick Hillman (2004) analyses how different understandings of justice are embedded in alternative approaches to river management in an Australian context. Traditional top-down engineered approaches have been challenged by newer catchment-framed, ecosystem-based approaches, and in this process alternative notions of what constitutes a fair distribution of resources and responsibilities for stream rehabilitation have become apparent. Determining what constitutes distributive fairness in this context is involved, having to deal with the contested causes of river degradation, substantial biophysical and social variability across catchments, and a range of disciplinary perspectives and knowledge systems. He identifies six ways in which distributive principles can be applied:

- *Equality of rights*: where resources for stream rehabilitation are spread thinly and evenly across an entire catchment, characterised as 'peanut butter management';
- *Utilitarian equality*: where the vision is to maximise overall catchment health, balancing costs and benefits in the allocation of resources, a homogenising approach;
- *Democratic equality*: where priority is given to the most disadvantaged biophysical or human parts of the system, or focusing on the 'worst bits';
- *Proportional equality*: where the status quo and historically derived priorities are maintained, also known as the 'grandfather principle';
- *Causal responsibility*: where the polluter pays principle is paramount – those that are deemed responsible for river degradation (industry and agriculture primarily) are required to fix the problem;
- *Merit based*: where resources go to those communities who are most active or who have the best performance in managing their rivers.

See Hillman (2005, 2006) for related discussions.

(Ikeme 2003), industrial pollution (Walker *et al.* 2005) and hazardous facility siting (Bryner 2002). In later chapters we will spend some time considering alternative approaches to determining justice in the distribution of waste sites, air quality, flood risk, greenspace, carbon emissions and climate change impacts. What constitutes justice in each of these cases, as we shall see, is not simply determined or readily agreed, and alternative perspectives can be quite reasonably sustained.

A number of these principles of distribution (along with the questions in Table 3.1) demonstrate that it is not only distribution of the direct environmental burden or benefit itself that can be at issue, but also other dimensions of distribution which interact with these. Three such dimensions are relevant and important for claims about environmental justice – vulnerability, need and responsibility.

Vulnerability: The notion of vulnerability captures the important point that not all people are necessarily equally affected by an equivalent environmental burden or able to cope with or recover from its impacts (Buckle 1998; Cutter *et al.* 2000). Physiological, social, economic and cultural factors may mean that an entirely equal distribution of exposure to a burden may still have very unequal impacts (Sexton 1996) – for example, children are more sensitive to various forms of pollution because of their higher metabolic and respiratory rates, the ongoing development of their nervous system and so on (see Chapter 5); older people are more susceptible to excess and unusual heat or cold (Brown and Walker 2008); people of different races can have different illness profiles for genetic reasons (Kuehn 1996); poorer people have far fewer resources to recover from disasters such as flooding; recent immigrants or others with language difficulties may struggle to understand environmental health or risk warnings and advice on how to protect themselves (Pulido 1994; Thrush *et al.* 2005). For such reasons multilayered claims of environmental injustice can be particularly powerful in showing how distributional inequalities in vulnerability compound distributional inequalities in exposure. Vulnerability runs as a particular theme through Chapters 5, 6 and 8.

Need: If we are focused on access to environmental resources, then need becomes important, as in some sense the opposite of vulnerability. Claims of justice may need therefore to be seeking not just a simple absolute equality, but one which reflects differentials in need. Some households may need more access to water than others, because of numbers of children or illnesses that require frequent bathing and cleaning. Older people can need better access to energy and heat than others to protect themselves from harm during cold weather. People living in stressful, high density neighbourhoods arguably need more access to greenspace for achieving calm and relaxation than others (see Chapter 7). Some communities may need better protection from flooding because they lack the resources to self-help, flood-proof their homes and/or recover from flood impacts (see Chapter 6).

Responsibility: In terms of responsibility, again the interaction between different patterns of distribution can be important. Hillman (2004) refers to 'causal responsibility' in his list of alternative justice principles (see Box 3.1 above), and the polluter pays and the challenges of deploying this principle are also discussed by Bryner (2002: 39) in his assessment of environmental justice claim-making. Whilst polluter pays recognises that the distribution of the production of a pollution burden (rather than the distribution of its consumption) can be the focus of fairness or justice concepts, it can also be the interrelation between the production and consumption of such burdens that is seen to constitute an unjust situation. Distinctions are made here between situations in which distributional inequalities both affect and are produced by the consequences of the actions or informed choices of the same people – the polluter is also the burden-taker – and those where there is a dislocation between those benefiting from and those suffering from patterns of distribution – the polluter and the burden-taker are distinct. Examples include, for urban air pollution, the contrasting distributions of those driving cars generating pollution and those suffering the health impacts of exposure to pollutants from car exhausts (see Chapter 5) and, for climate change, the contrasting distributions of greenhouse gas emissions and distributions of both exposure to changing regional weather patterns around the world and vulnerability to their impacts (see Chapter 8).

In these multidimensional, layered and interacting ways, concepts of distributive justice are undoubtedly central to environmental justice claim-making, and we will examine how they are in each of the subsequent chapters. However, to conceive justice purely as a matter of distribution is insufficient both theoretically and for capturing the nature of justice as practised and argued over by environmental justice advocates and other actors. To develop a more complete account we now need to include the interrelated notions of procedural justice and recognition.

Procedural justice

Procedural justice is widely seen as a necessary second concept of environmental justice to add to that of distribution. Low and Gleeson (1998: 24), for example, argue that 'attention must be paid not only to the substance of justice, justice of outcomes and consequences, but also to the justice of procedure'. Shrader-Frechette (2002) puts forward the Principle of Prima Facie Political Equality as the objective for correcting problems of environmental justice, combining distributive with participative justice. Drawing on Young (1990), she argues that a combined conceptualisation is needed because 'purely distributive paradigms tend to ignore the institutional contexts that influence or determine the distributions' (Shrader-Frechette 2002: 27). Both Torres (1994) and Schlosberg (2007) make a similar point in arguing that broad, inclusive and democratic decision-making procedures are a tool, or indeed a precondition, for achieving distributional justice.

As such authors make clear though, procedural injustice does not serve only as an explanation or cause of distributive injustice; if so, it would be positioned only in the process or the 'why things are how they are' element of Figure 3.1. It is also a subject or element of justice and claim-making in its own right, and therefore positioned as part of 'how things should be'. Many definitions of environmental justice convey the importance of fairness in procedure or process as a distinct concept of justice. Looking back to the definitions listed in Table 1.1 in Chapter 1, we find that they use phrases such as 'meaningful involvement', 'access to the decision-making process' and 'participation in decision-making' in identifying discrete objectives which constitute part of an overall environmental justice definition. Many accounts of conflicts erupting over a range of environmental concerns also emphasise the procedural as well as the material subjects of contestation. The literature on siting conflicts (see Chapter 4), in particular, has explored how exclusive and closed decision-making processes generate conflict over and above differences in evaluations of perceived environmental threats (Grimes 2005), with protest activity focusing on perceived injustices in procedure and the lack of opportunities to be heard or listened to (Boholm and Lofstedt 2005; Wolsink 2007).

Given the concern with decision-making processes, the practices of government and regulation are the principle focus of procedural justice claim-making – for Schlosberg (2007: 25) procedural justice comprises the 'fair and equitable institutional processes of a state'. However, other institutional settings and processes are also implicated, particularly given the 'roll out' of state functions under neoliberalism to non-state actors, including the private sector, public–private partnerships and third sector organisations (Heynen *et al.* 2007; McCarthy and Prudham 2004). A specific example is discussed in Chapter 7 concerning the 'shared governance' of urban greenspace. Indeed, through such processes there is clear potential for previously open processes of governmental decision-making to become hidden within the less public and less transparent practices of commercial or third sector organisations.

Wherever procedures and decisions are being enacted, we can approach claim-making for procedural justice using similar categories to those for distributive justice. A 'community of justice' – the recipients of procedural justice – has to be explicitly or implicitly defined in space and time. Often there is a particular focus on the people or community who are most affected by decision-making processes and the particular claim or right they have to be heard and included (Hunold and Young 1998; Shrader-Frechette 2002). If this is extended to include future generations (as with climate change – see Chapter 8), achieving their meaningful participation in decision-making is particularly problematic.

There are also various dimensions of procedure that can be the subject of justice and the focus of multidimensional injustice claims (Schlosberg 2007; Stephens *et al.* 2001). These include:

- the availability of environmental information, an essential condition of effective participation and informed consent;
- inclusion in environmental policy-making and decision-making processes in terms of who is allowed or enabled to participate, but also the resources available for participation, the equality of respect given to participants, and the degree to which power is shared and meaningful outcomes achieved (Bickerstaff and Walker 2005; Shulman *et al.* 2005; Zavestoski *et al.* 2006);

- access to legal processes for challenging decision-making and protecting environmental rights (McCracken and Jones 2003);
- inclusion in community-based participatory research in which scientists collaborate closely with community partners in the creation of knowledge about environmental concerns (Delemos 2006; Grineski 2006).

Hunold and Young (1998) include several of these dimensions within a detailed prescription for an ideal participatory democratic process for siting hazardous waste facilities – see Table 3.2. They argue that such a process is needed because it is unfair and ethically suspect to impose a risk on citizens without their participation in the siting process; that a fair process will mean the most distributively fair outcome is likely to result; and that the 'wisest' and most informed decision will also be made. Such ideal types are easily critiqued when compared to a reality in which unequal power relations and the authority to make decisions are not so readily put aside (Bickerstaff and Walker 2005; Tewdwr-Jones and Allmendinger 1998), but they do provide a set of conditions which claims for procedural justice might aspire to and work towards.

Lake (1995) argues though that to focus demands for effective participation and procedural justice only on isolated decision-making processes is insufficient. Drawing on Young's preoccupation with process and causation, he contends that where the environmental justice literature has gone beyond distributional justice

Table 3.2 Criteria for evaluating the degree of democracy of a siting process for a hazardous waste facility

Criterion	Conditions
Inclusiveness	Identifying all relevant groups in an area, facilitating their organisation, encouraging and supporting their involvement
Consultation over time	At all stages of the process; agenda setting, formulation, decision making, implementation and evaluation
Equal resources and access to information	Compensation of weaker parties for serious power disparities – informational or financial support; commissioning of 'counter studies'; translations into non-technical language
Shared decision-making authority	All those involved have a role in the final decisions of participatory process – a negotiated agreement or a vote
Authoritative decision-making	Decisions from the participatory process must be binding on public officials
Large unit of review	Site searching should be at state or regional level, allowing multiple communities to be considered and to make decisions
Nature of facility	The nature of the facility itself – size, purpose, standards – must be part of the discussion
Alternative methods	Alternative methods of dealing with the hazardous waste, other than one centralised facility, should be considered

Source: Summarised from Hunold and Young (1998).

to incorporate matters of procedure 'it adopts an unnecessarily truncated notion of procedural justice' (ibid.: 162), which fails to grapple with the need for local community control not only of the distribution of environmental problems but also of their production. As he states, 'arguments regarding procedures for obtaining a just distribution of environmental problems do not address procedures for deciding which environmental problems are produced' (ibid.: 164). Whilst in danger of overplaying the possibilities and politics of local action in the face of globalised flows of both capital and environmental bads, his intervention is important in demonstrating the limits of conceiving environmental justice solely as distributive and the need to move more deeply towards understanding causative processes (see later discussion of process dimensions). Further complexities involved in defining and making sense of procedural justice in practice will be discussed throughout later chapters, but in Chapter 4, in particular, where issues of self-determination and community consent in waste siting have to be worked through.

Justice as recognition

Whilst it has become commonplace to acknowledge that environmental justice claim-making is 'bivalent', incorporating both distribution and procedure, Schlosberg goes further to add a third concept of recognition. This, he argues, makes environmental justice 'trivalent' (Schlosberg 2004), involving issues of recognition distinct from but closely connected to those of distribution and procedure. In constructing this argument, he draws on justice theorists such as Young (1990), Fraser (1997, 1999) and Honneth (1995) who all argue, although in different ways, that misrecognition – in the form of insults, stigmatisation and devaluation – is fundamental to the damage and constraint that are inflicted on individuals and communities and to the production of distributional inequalities. As in the case of procedural justice, an integrated argument is made that sees recognition as both a subject and a condition of justice. It is a distinct, separate form and experience of injustice and terrain of struggle, but deeply tied to distributional inequalities.

At the core of misrecognition are cultural and institutional processes of disrespect which devalue some people in comparison to others, meaning that there are unequal patterns of recognition across social groups (defined by gender, race, religion, ethnicity and so on). Whilst the focus is in part on the institutions of the state which can explicitly or implicitly give unequal recognition to different social groups, and with potentially powerful consequences, there is a wider everyday cultural basis of misrecognition such that 'the conception of justice occupies social and cultural space beyond the bounds of the state' (Schlosberg 2007: 16). As Sayer (2005: 55) puts it, recognition is 'implicit in the way people address and deal with one another, whether they are kin, friends, associates or strangers, and in the merest looks of "civil attention" as Goffman termed it'. Fraser (1997) shows how social norms, languages and mores are fundamental to the failure to recognise and respect group difference and identifies a number of processes of misrecognition which lead to status injury. These include practices of cultural domination and oppression, being rendered invisible through non-recognition, and being routinely maligned or disparaged in stereotypical and stigmatising public and cultural representations.

The sense in which recognition is cultural and politically deeply embedded means that the specific and particular dimensions of recognition that are distinct to environmental justice claim-making can be hard to identify. Many definitions of environmental justice do not explicitly include notions of recognition, although Schlosberg argues that some do. He points to Bryant's (1995a: 6) definition, which states that: 'Environmental justice is served when people can realize their highest potential, without experiencing the "isms"', and makes reference to 'cultural norms' and 'values' that contribute to people interacting 'with confidence that their environment is safe, nurturing and productive'. He also argues that, both in the US context and in global justice movements, struggles over environmental justice are often centrally motivated by various forms of oppression and life experiences of disenfranchisement and discrimination, both intentional and structural in nature.

Whilst forms of racism are often emphasised, the scope of misrecognition extends far wider to include other dimensions of identity. Gender and disability have both, for example, been the focus of environmental justice activism and research and of related experiences of misrecognition. Charles and Thomas (2007: 218) focus on disability, arguing that 'the struggle for environmental justice must recognize the oppression of disabled people as part of the essential broadening of the notion of citizenship', with systematic processes at work that exclude disabled people as 'imagined citizens' and that render them invisible beyond a tick-box category. In a concern specifically with the deaf as a linguistic minority, they argue that administrative processes work with stereotypes and simplistic categories that at best reduce 'the disabled' to a single voice, hiding variety and difference, and that mainstream culture unconsciously isolates deaf people in everyday processes and interactions. Misrecognition with multiple dimensions is, in such ways, at the core of their concern for the environmental rights of disabled people and their invisibility in processes of environmental decision-making. Again, across subsequent chapters we will see where and how matters of recognition are raised – for example, in the concentration of risks in indigenous communities (Chapter 4), in the creation of and responses to flood disasters (Chapter 6) and in the demand for cultural survival in climate change debates (Chapter 8).

Integration and the capabilities framework

In moving through these three forms of justice concept, their mutual interrelation has been emphasised. For Schlosberg this is crucial; he persuasively demonstrates how all three concepts of justice work together, arguing that:

> These notions and experiences of injustice are not competing notions, nor are they contradictory or antithetical. Inequitable distribution, a lack of

recognition and limited participation all work to produce injustice and claims
for injustice.

(Schlosberg 2004: 529)

Indeed, he goes so far as to argue that, 'within the environmental justice movement, one simply cannot talk of one aspect of justice without it leading to another'
(ibid.: 73), a claim which may or may not be borne out across the diversity of
environmental justice framing and claim-making that goes on. Certainly there are
likely to be differences here between the overtly politicised claim-making of
environmental justice activists and that of (some) academics and government
policy bodies who utilise a more specific and restricted sense of justice and may
well resist certain forms of integration and combination. We will, however,
certainly see examples of their integration in working through the topics of subsequent chapters.

The observation that policy often works with a restrictive and insufficient
notion of justice links to the motivations at work behind the development of
the 'capabilities framework' by Sen (1999, 2009) as an alternative way of
approaching the definition of justice objectives. Schlosberg presents this as a
fourth category of justice concept. In some ways it is distinct from the others, but
it can also be thought of as an integrative framework within which various broad
understandings of justice – including distribution, procedure and recognition –
can be encompassed. The capability perspective (also referred to as a 'framework') has at its core a claim as to the appropriate 'space' (or informational
focus) for determining what justice should be. Sen's argument is that it is what
people achieve and are able to do that matters when making analyses of inequality and judgements of justice and injustice. His is an *accomplishment*-based
understanding of justice that 'cannot be indifferent to the lives that people can
actually live' (Sen 2009: 18). The important elements of what constitutes a good
and worthwhile life, the things that people value, are referred to as 'functionings'.
A person's capability to achieve these functionings is where the space for
determining justice is located, and functionings can take various forms:

> The concept of functionings ... reflects the various things a person may value
> doing or being. The valued functionings may vary from elementary ones,
> such as being adequately nourished and being free from avoidable disease, to
> very complex activities or person states such as being able to take part in the
> life of the community and having self respect.

(Sen 1999: 75)

There is a distinction made between functioning and capability to function, and
Sen illustrates this distinction in the example of the well-resourced person who
chooses to fast, as opposed to the famine victim who starves because he or she is
unable to access sufficient nutritious food. Both fail to achieve a key functioning
– nourishment – but one is exercising their choice not to eat even though they
have the capability to do so, whilst the other has neither the functioning nor the

capability to achieve that functioning (be able to access sufficient nutritious food to eat). Many things also structure capability or shape how income can be converted or translated into good and worthwhile lives and things that people have reason to value, including 'inborn circumstances ... as well as disparate acquired features, or the divergent effects of varying environmental surroundings' (Sen 2009: 66). In focusing on the processes through which distributed goods and resources are transformed into well-being and the leading of a sufficiently good life, both participation and recognition are incorporated into the framework. As Schlosberg (2007: 34) states, the capability approach focuses 'holistically on the importance of individuals functioning within a base of a minimal distribution of goods, social and political recognition, political participation and other capabilities'. Sen sees justice as having a plural grounding, as very different features of life had to be encompassed – human lives, he argues, need to be 'seen inclusively'.

There are relatively few cases of the capability perspective being applied within an environmental justice framing – see Tschakert (2009) for a recent example. It has though been more widely utilised in closely related analyses of inequality in access to environmental resources and vulnerability to the impacts of disasters in the developing world (see Chapter 6 on flooding), including Sen's work on famine. Schlosberg also identifies examples of environmental justice activism that can be interpreted as centring on both individual and community capability to function – campaigns against childhood asthma and threats to the cultural survival of Native American communities are two of these. Discourses of climate justice can also be focused on threats to capabilities to function, given the many interacting processes that are likely to come together to change the ability of people and communities to maintain their well-being around the world (see Chapter 8).

Evidence: how things are

The second central element of claim-making in Figure 3.1 is concerned not with justice per se, but with evidence: evidence of how things are, in particular, what is unequal and how this inequality is patterned and experienced by different social groups. It is the combination of evidence of inequality and claims about what makes justice and injustice that constitutes the core of the practice of environmental justice claim-making, wherever this is enacted. One without the other is undoubtedly less complete and typically less effective in providing a case for action, change or redress.

Two preliminary points are needed. First, the phrase 'evidence of inequality' could be taken to refer only to evidence of distributional patterns of burden and benefit. However, evidence on patterns of inequality in procedure (who has access and influence) and in recognition and misrecognition (who has respect, who is denigrated) is also very much relevant and included in what follows. The key is 'difference' and, as we have seen in the previous section, difference can be given normative meaning in various ways. The form that evidence takes and the type of justice concept that is applied are clearly connected; a concern for distributional justice tends to bring forward certain types of evidence, a concern for procedure or recognition, other types. But the relationship between justice concept and form of evidence is not fixed or absolute.

Second, and to reiterate a point made earlier, evidence about environmental inequality is not conceived here as unproblematic, a matter of simple fact and truth. Evidence is produced through social processes, with the attendant selections, contingencies and uncertainties this entails. Taking on board that the production of evidence is itself a form of claim-making is not a standard feature of much of the existing environmental justice literature – although Shrader-Frechette (2002: 194) is a notable exception, arguing that 'facts alone never determine all aspects of a situation. Facts are always incomplete and saddled with implicit interpretations.' Some discussions of environmental justice that pore in detail and in full critical mode over the minutiae of justice concepts are entirely accepting of the unproblematic 'facts' of the evidence. Maintaining critique is, however, necessary if the politics involved in the production, contestation and evaluation of evidence – which includes crucially the active denial of evidence claims – is to be revealed and understood; this will become an explicit theme of the later discussion. We can begin though by considering where evidence comes from and the various forms it takes.

Where does evidence come from and who is involved?

There is no one source of evidence of environmental inequality, no one authoritative and widely used metric or measure (as there can be for forms of economic and social inequality) and no routine or standardised requirement to analyse the social distribution of environmental impacts as part of decision making processes (Walker 2007). Rather, there is a constantly evolving patchwork of diverse pieces of evidence that are brought to bear within particular acts or processes of claim-making, a patchwork added to by a variety of actors for a diversity of reasons.

If we take the case of the UK, where evidence of environmental inequality has been explicitly accumulated over the past ten years or so (Lucas *et al.* 2004), the list of those producing evidence includes environmental activist and campaigning groups; academics and consultants; government departments and various authorities and agencies with environmental, health, local regeneration, planning and sustainability responsibilities. Within this 'new' and wide-ranging field of evidence generation, each of the actors has had different motivations for committing resources (time, effort, money, skills) to producing evidence and for focusing attention on particular dimensions of inequality – and therefore not on others. For example, Friends of the Earth (2000) undertook the first national-scale analysis of the relationship between industrial pollution and income in the UK (i) because it fitted with its ongoing campaigning work against industrial pollution and its developing social agenda; (ii) because pollution data was publicly available; and (iii) because the research was expected to generate results that could achieve an impact in the media. The Environment Agency undertook its own in-house analysis of similar data and then commissioned a group of academics (including me) to undertake a more thorough study (Walker *et al.* 2003), again for a set of reasons: (i) because industrial pollution was within its regulatory remit; (ii) because it was seeking to develop a more social dimension to its own work that would fit into the major policy concerns of the New Labour government; (iii) because there was a commitment amongst key individuals to making the organisation more sensitive to issues of inequality and justice.

In these cases (as in *all* others) decisions were taken about what was to be examined, in what depth, with what resources, at what scale and using which methodologies. Each of these decisions shaped the nature of the specific evidence about inequalities which was then produced, made available and publicised to others. Producing evidence is about committing effort, making selections and choices, and having the power and resources to do this.

Such contingencies and the awareness of power relations being at work within the production of evidence have led both activists and academics to call for and experiment with participatory, community-based research in which the training and expertise of researchers are set within a collaborative process with local people (Delemos 2006). Indeed, as noted earlier, such involvement has been included as a necessary principle of procedural justice in its own right. More specific cases and examples of community participation in evidence generation are considered in Chapter 5 on air quality.

Whilst such participatory processes focus on acts of deliberate research and data collection, evidence of environmental inequality is also accumulated experientially as part of everyday life or through people being involved in particular activities or disputes. This brings evidence very much within the realm of the lay person or citizen who can make observations and bring their own expertise and experience to bear in making claims about injustice, unfair treatment, misrecognition and so on. This emphasises the need to think about both justice *and* evidence in broad terms, and being open to different epistemologies and forms of knowledge (Irwin *et al.* 1999). Indeed, Shrader-Frechette (2002) makes the case for all citizens having an ethical responsibility to engage in environmental justice advocacy rather than leaving this to 'professionals' of various types – an ethical responsibility that is needed to counteract the power of vested interests and the tilting of the playing field to their advantage.

Evidence as statistical quantitative analysis

Given that distributive concepts of justice have been dominant in much of the framing of environmental justice concerns, it is no surprise that much of the evidence of environmental inequality that has been produced and given political significance focuses on distribution and takes a statistical, quantitative form. Establishing who gets what in terms of benefits and burdens, who is vulnerable and who has responsibility calls for some form of measurement, analysis and comparison. Doing that in quantitative terms is generally seen as more authoritative and reliable than relying on presumption or impressionistic assessments.

Accordingly, there are a whole host of studies that report on statistical patterns, correlations and regressions which demonstrate the extent to which environmental phenonema are equally or unequally distributed across particular defined social groups. Many examples of studies focusing on the quantitative analysis of patterns of waste sites, air quality, flood risk and greenspace provision – and more besides – will be examined in later chapters.

As noted in Chapter 2, studies of this form were particularly influential in the early emergence and shaping of the environmental justice movement in the US, focusing predominantly on using data on the locations of polluting installations or on emissions or levels of pollution and relating this to data on ethnicity and social class. Statistical studies have increasingly focused on other environmental phenomena and other forms of socio-demographic difference (such as age, health status, housing tenure), reflecting the broadening of the scope of environmental justice concerns. Analysis has been undertaken at different scales (global to very local) and for different time periods (the current day, forward into the future and back into the past) and focused on a range of different places around the world. The spread of such studies and the coverage of different topics within this growing evidence base reflect the emergence of particular framings and priorities in contexts of time and place.

However, there are also more practical considerations that shape the patchwork of where evidence has and has not been produced. The availability of data is fundamental, as statistical studies typically work with large environmental and socio-demographic datasets gathered and compiled by government agencies. Early environmental justice researchers in the US were able to take advantage of reasonable quality official data sources and rights of access to information which enabled actors outside of government to undertake their own bespoke analyses. Without such data access the classic, frequently-cited studies, such as that undertaken by the United Church of Christ on the distribution of hazardous waste sites in 1987 (see Chapter 4), could not have been produced, and evidence supporting claims that local struggles were part of a broader picture of environmental racism would not have been in place. As access to information rights have been extended through EU directives and the Aarhuus Convention (McCracken and Jones 2003), an improving environment for undertaking such studies is also emerging in Europe, if limited by the quality of the environmental data that is collected and compiled by the many governmental agencies involved. In many parts of the developing world, though, the possibility of undertaking *any* analysis of already existing official data is severely constrained by the limitations of both the environmental and the socio-demographic data sets. If useful official data is collected at all, it can be unreliable and incomplete, potentially entirely missing some of the more marginal people and places that may be of most concern.

Even where data is available, it is rarely in an ideal form and entirely 'fit for purpose'. In particular for environmental data it can be necessary to use proxies or surrogates (Buzzelli 2007) which may only partially capture what is really of interest – and proxies generate uncertainties of varying magnitudes. For example, in the case of the social distribution of flood impacts (see Chapter 6) a 'post-flood' analysis may be able to work with data that directly corresponds to what is at issue – i.e. who died in the New Orleans flood, whose houses were flooded, whose weren't. Such data is measuring the social distribution of actual proven impacts, even if these may not be entirely reliably recorded or uncontroversial in their measurement and categorisation. A 'pre-flood' analysis of who is at risk of flooding has to work not with proven impacts but with potential ones, relying on records of past floods or hydrological models of flood extent to estimate current or future levels of risk of a flood happening in any location and assumed or estimated impacts on people and property. If such modelling and mapping is not available, other more basic proxies may need to be used – Ueland and Warf (2006), for example, use elevation above sea level generated from a digital evaluation model as a crude indicator. If the impact of climate change on patterns of flooding is then factored into the equation, as in a 'foresight' study of future flood risk in the UK (Evans *et al.* 2004), uncertainties around where and how severely flooding will be experienced are further compounded. Knowing that evidence has uncertainties attached to it does not make that evidence worthless – far from it – but the limits of what is being claimed as factual evidence need to be recognised and evaluated in reaching conclusions and drawing out implications.

The use of proxies is only one of a series of methodological complexities involved in undertaking a statistical analysis of environmental inequalities. Many studies analyse distributions in spatial terms, a geographic approach using the technology of Geographical Information Systems (GIS). This conceptualises the impact of the environment on health and well-being typically in terms of where people live – and, in some studies, where children go to school (Sexton *et al.* 2000). This brings into play complexities related to the adequacy and comparability of the areal units for which data is available, the type of spatial relation that is conceived between people and environmental phenomena, and the scale at which analysis is undertaken (Liu 2001). These and other methodological complexities are summarised in Table 3.3.

As an example of how such complexities can play out in constructing a research design, Box 3.2 discusses the case of a study of inequalities relating to river water quality that I was involved in (Damery *et al.* 2008b) and was commissioned by the Environment Agency (see also the discussion of a greenspace study in Chapter 7).

The fact that such complexities exist and that many methodological decisions have to be taken means there is great scope for critique and disagreement about what constitutes a 'good' or sufficiently robust research design (however those criteria are defined) for distributional studies of patterns of environmental inequality. Retrospective review has become increasingly common, particularly of the body of statistical studies that was produced in the 1980s and 1990s in the US and became so influential in shaping policy (Brown 1995; Weinberg 1998). Bowen (2002) undertook a systematic analysis of 42 studies. Taking a strictly positivist position and basing his evaluation on 'reasonable scientific standards',

Table 3.3 Methodological complexities in GIS-based environmental inequality studies

Category	Explanation
Selection of study population	What is to be the population under study, in terms of its relevant socio-demographic or cultural characteristics, and its boundaries?
Impact on health or well-being	What is the assumed relationship between environmental parameter and good or bad impacts on people living in particular areas?
Data availability and quality	What environmental and social data is available, how good is the coverage, how is it sampled, what reliability issues are there, does it directly measure what is of concern or provide only a proxy or surrogate?
Spatial analysis	What spatial units are to be used for aggregating social data and by what method(s) is the association between people and environmental phenomenon to be analysed?
Comparison areas	What areas are to be used to make comparisons that determine the extent to which patterns and associations are significantly different or disproportionate?
Statistical methods	What methods are to be used to establish the statistical significance of patterns of spatial association between social and environmental variables?

Source: Adapted from Mitchell and Walker (2007).

Box 3.2 River water quality and deprivation: a problematic research design

This study was part of a wider project and was driven not by particular expressed concerns over the justice implications of river water quality, but by the fact that the EA had datasets and was responsible for the regulation and management of the water environment. The analysis was exploratory, therefore, seeing what patterns of relationship between river water quality and the degree of deprivation of nearby populations might emerge. In following through this brief, the first task was to establish the relevant relationship between river water quality and impacts on health or well-being, as this would shape other aspects of the methodology.

Complexities immediately arose. Were there any clear or potential health impacts from river water on nearby residents? Only from getting into or directly drinking the river water, and this required active involvement not just 'living nearby'. Or maybe, at a stretch, there could be neighbourhood risks from rats infesting heavily polluted rivers? The alternative was to see the river and its corridor not as a bad but as a good – the blue equivalent of greenspace – so that the concern became the 'absence of a good' that would otherwise be there to be enjoyed by those living nearby if the water quality was better. But then the chemical and biological data on river water quality that was available didn't really tell us anything at all directly about the quality of the river corridor in terms of its visual or recreational value, or indeed if the river was accessible to being seen or used in this way. And nor did the data tell us anything about the size or scale of the river, in itself a problem in analysing the deprivation characteristics of 'nearby' populations: where should nearby be measured from – the river bank or the middle of the channel? And how far should 'nearby' extend – 400, 600 or 1,000 metres? And on what basis should that distance be set?

Each of these questions was resolved to enable analysis to take place, but only in ways that could not fail to reveal the severe limitations on the validity and usefulness of the results that were produced.

he divides the reviewed studies into three categories – high quality, medium quality and poor quality. His evaluation of the latter is scathing:

> The studies categorized as poor quality have substantial enough flaws to be judged useless in terms of contributing anything to scientific knowledge, and, in the author's view, the study conclusions should therefore not be considered as having any merit whatsoever for public policy and management decisions. This category also includes research that was not designed, conducted, or documented well enough to know for sure whether it contains *any* scientifically meritorious conclusions, and so should be considered purely conjectural.
>
> (ibid.: 3)

The studies falling into the medium or high-quality categories are seen as more scientifically robust, and, since this assessment continued methodological advances have been made. More sophisticated evidence production has, for example, better addressed the impact of changing spatial units of analysis and making population assignment choices (Most *et al.* 2004), made better use of data on exposure and health outcomes rather than relying on problematic proxies and surrogates (Buzzelli 2007), and taken forward cumulative rather than single-topic risk assessments – see examples in Box 3.3. Further discussion of such methodological developments can be found in Chapters 4 and 5 in relation to patterns of waste siting and air pollution.

Box 3.3 Environmental justice studies analysing cumulative risks

A range of approaches for assessing multiple or cumulative risks or impacts have been developed in the environmental justice literature. Some of these restrict their aggregations to sets of related phenomena. For example, Lejano *et al.* (2002) focus on multiple toxic exposures through air pollution to calculate cumulative cancer risks and cumulative non-cancer hazard indices in Los Angeles. Relating the spatial distribution of these cumulative indicators to socio-demographic data, they find associations between high cancer risks and low income and minority (particularly Latino) communities. See Su *et al.* (2009) for a later more sophisticated analysis again for Los Angeles. Walker *et al.* (2003) calculate a combined 'air quality index' for England and Wales, aggregating together data for four different measures of air quality to identify a dozen pollution-poverty hotspots where poor air quality and high deprivation coincide, finding large clusters in parts of London, Manchester, Sheffield, Nottingham and Liverpool and small clusters elsewhere.

Bolin *et al.* (2002) go further to combine data on the locations of four types of hazardous industries and contamination sites in Phoenix, Arizona, to create a 'cumulative hazard density index' which measures the agglomeration of hazard zones within a census tract. Krieg and Faber (2004) go further still to bring together a set of 17 datasets of 'ecological hazards' for Massachusetts which range from hazardous waste sites, to sludge landfills, trash transfer stations and power plants. They give each of these sites a points rating to indicate the severity of each form of hazard, ranging from 1 through to 25, and aggregate scores for each town and city. Sicotte (2010) provides a very similar analysis for Philadelphia.

Fairburn *et al.* (2009) include a greater diversity of risks, combining air quality, industrial and waste sites and flood risks to produce an 'impact intensity score' that relates to the number of households within each census area that are affected by each form of risk. Applying this to South Yorkshire in the UK, they find a strong skew of high scores towards the most deprived deciles. The most advanced study to date is that by Pearce *et al.* (2010) for the UK, who construct a 'multiple environmental deprivation index' for census wards out of variables that cover air quality, temperature, UVB index, proximity to industry and greenspace availability, and then relate this to a health indicator (all cause mortality). They find that as the index value rose, so did income deprivation, and health also worsened even after taking account of the age, sex and socio-economic profile of each area.

Many academics and other actors have remained firmly wedded to improving the established tools of positivistic scientific inquiry and strongly supportive of the role for advanced epidemiological and toxicological evidence in progressing claims for environmental justice (Dutcher *et al.* 2007; Sexton and Adgate 1999). However, some critiques have challenged the ways in which scientific methods have been applied. Foreman (1998) recognises the weaknesses at the core of the US evidence base but argues that the classic empirical studies had political rather than scientific significance, and in this context 'formal analysis is to a considerable extent irrelevant to the underlying objectives and gratifications that stir activist and community enthusiasm for environmental justice' (ibid.: 27). Emphasising the practical politics involved in US environmental justice activism, he sees this as challenging technocratic ways of thinking that rely on the tools of positivist methodology to the point that 'activists and angry community residents are disinclined to allow epidemiologists, toxicologists and statisticians to define the premises of their movement' (ibid.: 29).

Bryant (1995b) makes similar arguments in calling for participatory approaches to research design which can at least place the tools of scientific analysis within a less top-down and expert-determined process. For some though the use of risk assessment is more fundamentally problematic, with Kuehn (1996), for example, claiming that risk models assume an 'average white male', failing to take account of the different sensitivities than can exist amongst other body types. Such assertions prompted Simon (2000) to mount a trenchant defence of risk assessment methods, claiming that, 'like a hammer, risk assessment is merely a tool and has no moral value on its own' (ibid.: 559) and accusing environmental justice advocates of 'neo-Luddite' tendencies. As will be explored further later, such exchanges clearly demonstrate that what counts as useful and valid evidence of environmental inequalities is not unproblematic or uncontested and is as much open to the politics of environmental justice as other elements of claim-making.

Evidence as qualitative and experiential

Not all of the dimensions of inequality that are relevant to a pluralistic understanding of environmental justice can be adequately captured through quantitative measures. Conceiving justice as procedure and recognition, in particular, moves us towards other forms of evidence which rely less on the analysis of large-scale data sets and more on particular cases, experiences and narratives, captured qualitatively through the accounts and observations of those involved. Whilst some elements of procedural fairness may be recorded quantitatively (such as counts of who is involved at meetings or discussion fora), gaining a view of the barriers and inequalities involved in accessing information, in resourcing involvement or having influence needs to draw on 'thicker', more multidimensional accounts. For example, Bickerstaff and Walker (2005) use interviews with people involved in participatory processes for the development of local transport plans in England to reveal the ways in which strategic behaviour and tensions around the purpose and outcomes of participation were felt and experienced by those involved. Hunold and Young (1998) use their own observation and monitoring of a siting process in Switzerland, and the accounts of participants, to assess the extent to which it satisfied their normative ideal of a fair process.

In exploring the relevance of misrecognition to environmental justice activism, Schlosberg (2007) recounts various examples of exchanges in meetings and public hearings in which the subtleties of respect reflected in modes of address were significant in crediting the knowledge of some participants and discrediting that of others. Kurtz (2007: 417) found similar processes at work in how state officials disrespected the members of an environmental justice protest group she was researching, referring to them as 'hysterical housewives' and 'a bunch of gray-haired ladies'. Such qualitative forms of evidence are also needed to understand how mechanisms of stigmatisation of both people and places can rest upon the subtle inflections of everyday exchanges and encounters – the way that humour is used and directed, the associations that are made on telling someone where you live and so on (Simmons and Walker 2005).

Qualitative methods have also been used to reveal the values and subjectivities that underpin how different environmental goods and bads are understood by different social groups. Brownlow (2006a), for example, in work discussed further in Chapter 7, uses the narratives of residents, government and park officials gathered from interviews and focus groups to understand how patterns of control of and access to public spaces had changed over time for different population groups and how associations of neglect, disorder and risk had become connected with particular cases of urban nature. Only through methods that generate narratives could the complexities of why proximate greenspace may not be valued or utilised as an environmental good, and the inequalities that this reflected and produced, be understood. Day (2010) uses 45 in-depth interviews to understand the subjective relation between older people and their local environments, revealing processes of exclusion from and within their urban environments and narratives that ranged across different justice concepts. Tschakert (2009) uses qualitative participatory methods in his research on environmental justice in the artisanal gold-mining sector in Ghana in order to enable the miners to 'articulate their own definitions of desires, capabilities and flourishing' (ibid.: 714), as well as to reveal the many different ways in which these 'illegal' workers were misrecognised. These included violation of the body through torture and assault, denial of rights to political activity and participation, and denigration of ways of life through negative stereotypical representations in the media and in the statements of government and industry officials.

Participatory methods have also been argued to improve the quality of distributional analysis that can be undertaken, with interactions between different forms of knowledge. Lay epidemiology, carried out by local people and drawing on their knowledge of patterns of illness and disease rather than only on collected health statistics, has been a demonstrably powerful approach in particular cases. Corburn (2002) reports on the carrying out of a cumulative risk assessment which draws on local experiential knowledge to identify sources of local pollutants and inform the assessment of patterns of exposure. He argues that the use of interviews, primary texts and ethnographic fieldwork can improve local policy through adding to the knowledge base and bringing in new and previously silenced voices (see Chapter 5 for a related discussion on air-quality evidence collection).

The politics of evidence

The implication of much that has already been said about the diversity of forms and sources of evidence, and the methodological complexities inherent to its production, is that evidence of 'how things are' has a politics, just as do other dimensions of claim-making. As Gieryn (1983, 1999) makes clear, a claim for knowledge is a claim for authority and power, and there can be active and strategic 'boundary work' going on around drawing the line between what is good knowledge and to be respected and what is inferior or bad and therefore to be discounted. Alternative knowledge claims about patterns of environmental inequality can be in competition and subject to critique, like other dimensions of environmental knowledge controversies. An example discussed in Chapter 4 concerning disagreement between researchers in the US over analyses of patterns of waste site locations will more than ably demonstrate this. Other cases have been apparent in the discussion up to this point. For example, the denigration of women, people of colour or indigenous groups in public hearings and meetings is also a rejection of the evidence they have to present, which may have quite different bases to those of official, establishment sources. The accusation that evidence of discrimination is 'largely anecdotal' and therefore of no worth is identified by Shrader-Frechette (2002: 13) as a recurrent tactic of environmental justice denial.

Public hearings and meetings are examples of the many forms of arena in which evidence claims are subject to critique and evaluation. Others include various forms of legal and political process, academic journals and crucially the media. Whose evidence counts in such arenas is intrinsically wrapped up with inequalities in recognition, participation and distribution, providing a further way in which justice and evidence are intertwined. Those that have respect, are able to participate and have the resources to collect, analyse and present evidence are able to make their knowledge and knowledge systems count in the way that others are not. Whilst some would like to represent environmental justice activists as having privileged access to the media through using 'unaccountable populist scare tactics' (Friedman 1989), critiques of how establishment science and public debate has been captured and corrupted by vested corporate interests are more persuasive. Faber (2008), for example, documents at length how 'think tanks' heavily funded by corporate bodies have increasingly dominated public debate in the US, providing 'expert' testimony and commentary that claims independence and reliability but, in reality, has neither of these qualities. He argues that, through working with industry 'front groups', lobbyists and public relations firms, corporate interests have increasingly exerted ideological control over the production and evaluation of environmental scientific knowledge. Hence they also exert control over state bureaucracies that draw on such knowledge, the decisions they make and the policies they enact. Shrader-Frechette (2002: 186) similarly sees a 'tilted playing field created by the unequal power of vested interests' as unfairly distorting the production and evaluation of evidence and the making of claims about environmental justice concerns.

Process: why things are how they are

The third element of claim-making, identified in Figure 3.1 as linked to both justice and evidence claims, is concerned with process. Claim-making about process seeks to explain the causes of environmental inequality and injustice, how patterns of inequality are produced and reproduced, and why some in society suffer the downside and injustice of the intertwining of environment and social difference whilst others do not. A focus on process has been repeatedly seen as lacking or insufficiently developed in much environmental justice scholarship. For example, in the late 1990s Weinberg (1998: 613) argued that the literature has too often 'been at a loss to explain the results of the sophisticated analysis', seeing this not just as a gap in academic endeavour but also as having consequences for policy: 'without a better understanding of causal connections, it is hard to imagine being able to make any meaningful policy prescriptions'. More recently, Schweitzer and Stevenson (2007) have similarly criticised the lack of engagement with explanatory theory in much environmental justice research.

Recent academic work has to a degree addressed this deficit, in part by tying environmental justice to existing bodies of social and political theory. Looking more broadly across environmental justice discourse, we can also find a variety of claims about process and causation of varying depth and robustness. In simple terms, we can distinguish between process claims that are contextual and situated and those that are structural:

- *Contextual process claims* begin with specific situations, for example, explaining why a pattern of poor air quality in a city is concentrated on poor and minority populations, which might typically involve tracing local patterns of urban and industrial development and decision-making over time.
- *Structural process claims* begin from a perspective on how society works, how power is distributed and how uneven environmental outcomes are the consequence of the systemic structuring of social relations, and so on.

In practice, we can find much movement between the contextual and the structural in claim-making, with the details of particular cases being used to exemplify and explore the working out of broader processes, and vice versa. In later chapters we will examine examples of contextual process claims articulated around, for example, particular cases of waste facility siting, particular flood events and specific forms of climate change impact. In this section, therefore, the focus is on various accounts of structural processes that have applied to environmental justice concerns. The discussion draws primarily on academic literature and perspectives that set out theoretical frameworks for understanding why inequalities (of various forms) are how they are, but the relation between these and the claim-making of other actors is discussed in the final sub-section.

Before going further though, it is important to note that we have already encountered process claims about the production of inequalities and injustice. In discussing distribution, participation and recognition, we found connections between them, and a degree of overlap between justice concepts and questions of process. Schlosberg's pluralistic position on the definition of environmental justice contends that, as well as distribution, participation and recognition being distinct forms of justice in their own right, each can also explain the existence of injustice in the others – they interact and are mutually constitutive. As he nicely observes:

> if you are not recognized you do not participate; if you do not participate you are not recognized. In this respect justice must focus on the political process as a way to address both the inequitable distribution of social goods and the conditions undermining social recognition.
>
> (Schlosberg 2007: 26)

This key observation can be extended, as in Figure 3.2, into a whole set of inter-relationships in which each form of injustice is explained (in part) through its linkage to the others.

This is a useful starting point and identifies forms of interrelation that frequently figure in the claim-making of environmental justice activists. But on their own these connections do not cover all of the explanatory claims that are made or capture the range of languages and frameworks that have been utilised.

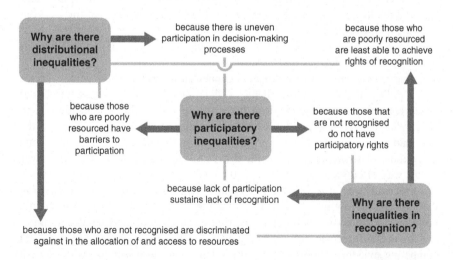

Figure 3.2 Explanatory interrelations between distribution, participation and recognition.

There are others to consider, beginning with claims using the language of environmental discrimination and racism.

Environmental discrimination

The emphasis on forms of discrimination as an explanation for patterns of environmental inequality reflects, in part, the importance of race in the development of the environmental justice movement. As we saw in Chapter 2, in the US, and also in other countries such as South Africa, the argument that discrimination produces the disproportionate burdening of minority racial, ethnic or indigenous groups became a powerful part of the political framing of environmental justice. Looking beyond this, we can also see other forms of discrimination, such as by gender, disability and age, increasingly being enrolled into the process claim-making of environmental justice activism and research. As we shall see though, there is much contestation over exactly how environmental discrimination of various forms is to be understood, its significance and its interplay with other processes producing uneven environmental outcomes (Kurtz 2009).

The focus on racism as an explanation for unequal patterns of exposure to environmental bads has been extensively examined and debated in the US in particular. The first use of the term environmental racism is generally attributed to Benjamin Chavis, the former head of the United Church of Christ's Commission on Racial Justice (Holifield 2001). He defines it in relation to multiple forms of act and outcome relating to policy, siting and decision-making processes:

> Environmental racism is racial discrimination in environmental policy-making and enforcement of regulations and laws, the deliberate targeting of communities of color for toxic waste facilities, the official sanctioning of the presence of life-threatening poisons and pollutants for communities of color and the history of excluding people of color from leadership of the environmental movement.
>
> (Chavis 1994: xii)

The charge of environmental racism is connected here to multiple actors. To the state ('policy-making', 'enforcement', 'official sanctioning'), to those 'targeting' waste facilities, which could mean both private and public organisations, and to the leadership of the established environmental movement, which at the time had a predominantly white profile and, it was argued, predominantly white and middle-class preoccupations (Field 1997). Whilst environmental racism was in this way often seen as multidimensional and operating through multiple societal processes (see also Mohai and Bryant 1992), for a period debate focused in the US on a narrow question of discriminatory intent in the siting of waste and polluting facilities (the 'deliberate targeting' dimension of the Chavis definition) and in the enforcement of environmental regulations. The ins and outs and competing evidence claims of this siting debate are discussed at greater length in Chapter 4.

In South Africa under the apartheid regime it was clear that racism was intentional, explicitly embedded in legislation and the institutions of the state with consequences for the patterns of environmental quality and access to environmental resources experienced by different racial groups. However, to see racism purely in these terms, as something necessarily explicit and intentional and restricted to overtly racist regimes, is to ignore its more subtle institutional forms and legacies. Writing from a US perspective, Bullard (1990: 98) has argued that the 'roots of institutional racism are deep and have been difficult to eliminate', setting contemporary experiences in the context of the history of racial segregation in the US which continues to shape the contrasting fortunes of white and non-white populations. He sees environmental racism as being reinforced by 'governmental, legal, economic, political and military institutions' through embedded and normalised processes that lead to 'benefits for whites' while shifting costs to people of colour (ibid.: 98). An example, he argues, is how land-use zoning policies are used to direct industry towards particular areas that are historically already industrialised, or where vacant and cheap land is available. Whilst a normal and accepted part of planning logic, such zoning policies also have the consequence of directing industry towards places often populated by racial minorities. Other mechanisms through which he sees communities of colour having fewer choices and less influence over decision-making include the lack of representation in key environmental bodies such as the EPA, regulatory boards and commissions; biases in federal housing policies and the operation of housing markets; and high levels of unemployment in black communities and the need therefore to take low-paid and health-threatening jobs (note here the causal connections being made between different forms of injustice, as in Figure 3.2).

Pulido (1996) takes a critical perspective on debates about environmental racism and identifies three pitfalls with the dominant perspectives (see also the discussion in Kurtz 2009): first, that racism is contained, a specific thing whose effects can be isolated from others; second, that racism is intentional rather than being ideological such that it 'infuses society, including culture, politics and economic structures' (Pulido 1996: 144); third, that racism is monolithic, unitary and fixed, without the recognition that many different racisms can co-exist and that 'non-white communities' can be fractured in other ways, such as by class and gender. Her critique therefore goes further than the intentional–institutional racism distinction, to question the primacy of racism as a sufficient explanation of process. In this respect she is critical not only of those who seek to deny the existence of racism unless absolutely proven through statistical evidence, but also of Bullard and others who follow universalising narratives of environmental racism that can serve to 'silence other visions, interpretations and experiences' (ibid.: 156).

The need to look beyond solely race-focused accounts and forms of discrimination becomes more apparent when other dimensions of difference and identity are considered. These are not as well developed in the environmental justice literature, although some more substantial attention has been given to gender. Buckingham and Kulcur (2009) argue that gender has been neglected in environmental justice research in part due to the less publicly visible scales at which gender differences are felt – primarily those of the individual, the body and the household – and the ways that women's family and social roles are viewed and defined. For example, women's bodies can be particularly sensitive to various forms of pollution; women typically have to manage most directly illness amongst children and other family members that comes from being exposed to harmful pollutants (Sze 2004) and other forms of hazard; they can suffer disproportionately from the stress of high-density urban living and from the consequences of unreliable access to key domestic environmental resources; and they are typically held most responsible for reforming household practices such as waste management (Buckingham *et al.* 2005). Buckingham and Kulcur (2009) identify a range of processes that have marginalised the identification of such environmental-related gender inequalities and thereby served to sustain their production. These include institutions (state and non-state) at a range of scales which are structured by gender inequalities; campaigning organisations that are 'gender blind at best, masculinist at worst' (ibid.: 661); an academic community which marginalises the study of gender inequality; sexism in the media and the use of sexist and derogatory language in participatory processes. Even though women have been key actors in the environmental justice movement in the US (Dichiro 1992), and in other grassroots environmental organisations around the world, various forms of discrimination have therefore limited their influence, participation and recognition as valued voices with particular interests and concerns.

Capitalism, class and the polluter-industrial complex

A second body of explanation of process begins not with a dimension of identity and associated claims of discrimination, but with frameworks rooted in understandings of how capitalism works to produce uneven environmental outcomes. The roots of environmental inequalities in these accounts are therefore to be found in political-economic processes, the working of capitalist markets and institutions, and relations between classes.

Whilst early foundations can readily be found in Marxist and related writing, one of the first theoretical accounts explicitly to be tied to environmental justice concerns is Schnaiberg's 'treadmill of production' (Schnaiberg 1980). The treadmill theory sought to explain why the expansion of the US economy was accompanied by rapid environmental degradation. Schnaiberg argued that in the post-war period capital was being accumulated at ever higher levels and being used to invest in various forms of technology that replaced labour-intensive processes. These technologies demanded more natural resources and produced more pollution and waste, and to keep profits rising more and more production, investment in technology and greater returns on this investment were needed. Employers, the state and unions all supported such ongoing investment as the way to achieve growth and replace the jobs lost to technology, whilst the consequences for environmental resource depletion and pollution were down-played. This continual treadmill of seeking an increasing return on capital invest-ment, downsizing of labour and consequent environmental degradation was accompanied by growing social-class segregation of the population, with middle-class workers and the professional classes who benefited from the treadmill moving to the suburbs away from the pollution and waste, thereby becoming distanced from the consequences of the investment from which they prospered – see Field (1997) for a closely related account of the unequal processes of urban-industrial change.

Treadmill theory is characterised by Gould *et al.* (2004) as primarily an economic change theory, focused on decision-making in the realm of production (rather than in the 'demands' of consumers) and emphasising the power inequal-ities that put political and economic elites in much stronger command of social forces than others who seek to mount resistance, including to socio-environmen-tal consequences. They argue this means that, despite challenges from labour unions, environmental groups and social justice activists at local through to global levels, the treadmill structures have rapidly adapted to these. Looking back over the preceding two decades, they conclude that:

> We could state boldly that *increasing the return on investment has displaced every other social and environmental goal* in this period. Moreover, this prin-ciple has become dominant in more societies through the forms of globaliza-tion that have been dominated by investors from the previously industrial societies. Indeed, this principle is increasingly dominating all forms of glo-balization, despite the resistance by socially and environmentally progressive forces.
>
> (ibid.: 305)

Pellow is one of several authors integrating the foundation provided by tread-mill theory into a specifically environmental justice framing, both in relation to the empirical investigation of forms of environmental injustice and in developing his own account of a process-based understanding of 'environmental inequality formation'. Pellow (2000) stresses the need to understand environmental inequal-ities as historically produced and as an evolving socio-political process rather than a discrete event in which victims are imposed on by perpetrators. He focuses on relations between stakeholders in arguing that environmental inequalities derive from an unequal struggle between stakeholders competing for scarce resources within the political economy. Those who are unable to mobilise resources suffer from environmental injustice; those with greatest access are able to deprive other stakeholders of that access. Key scarce resources he lists as power, wealth and status, along with clean and safe living, working and recrea-tional environments. Through this framework he sees environmental inequalities not as being simply and directly imposed by one class on another, but as emerg-ing from a process of continual negotiation and conflict between stakeholders. This pluralistic framework therefore enables a 'way in' for activists as part of the array of competing stakeholders, rather than activism being entirely dominated by the overarching structural forces of capitalism.

Daniel Faber (2008), in a rich and extended analysis, also provides a complex view of the processes through which environmental injustices are produced. He emphasises the need to reflect the contemporary conditions of globalised capitalism, in this way moving on from the 'post-war' focus of treadmill theory, but, as with Schnaiberg and Pellow, he still approaches his analysis from a US perspective. He integrates four key claims into his framework:

- In order to compete in the world market American business has a first imperative to become more efficient, reducing production costs through closing unprofitable businesses, cutting costs associated with environmental protection, reallocating money from established businesses towards venture capital in new and often more damaging activities, and relocating operations to pollution havens overseas.
- The state is being colonised by the 'polluter-industrial complex' with a class war being waged against popular social movements such as that for environmental justice. The corporate elites that benefit from a weak regime of environmental regulation are wielding power over the state apparatus through various means, including financing business-friendly politicians, employing corporate lobbyists and corrupting independent scientific investigation.
- Neoliberal reforms under the Bush administration have rolled back key elements of environmental and social regulation, cutting staff and budgets in the Environmental Protection Agency and other key bodies. Free-market forms of neoliberal environmental policy are being introduced which increase corporate flexibility to meet weakened environmental regulations whilst continuing to pollute in a profitable manner and burdening poor and marginalised communities far more than others.
- The export of ecological hazard from the US to less developed countries (see Chapter 4) is taking place through a suite of processes that include foreign direct investment in domestically owned hazardous industries, investment schemes to gain access to new environmental resources, the marketing of more profitable but also more dangerous products into emerging economies, the dumping of wastes and the relocation of polluting production processes. Corporate-led globalisation is thus serving to 'magnify externally and internally based environmental injustices to the advantage of the US' (p. 174).

In combination, he argues, these constitute a powerful set of interacting and mutually supporting processes that are continuing to sustain, strengthen and produce new instances of environmental injustice around the world. Whilst global in scope, Faber's and Pellow's analyses focus on the production of environmental injustice primarily in terms of forms of waste and pollution and in relation to US corporate business interests, which in both respects limits their wider applicability.

Urban political ecology

A further explanatory framework, applied more readily across a diversity of socio-environmental concerns, is provided by political ecology, and specifically urban political ecology. Political ecology as a field of academic scholarship emerged from work in less developed countries in the 1970s and 1980s engaged with questions of access to and control over land-based resources and in understanding processes leading to land degradation, soil erosion, water pollution, deforestation and similar environmental problems (Watts and Peet 2004). Political ecology at this time was distinctive in asserting that environmental problems were social in origin and definition, shaped primarily by political and economic forces. Powerful explications of the mutually causal relationships between marginality, poverty and environmental degradation were generated, with multilayered analyses setting local practices and ecologies within wider regional and global processes (Blaikie and Brookfield 1987). Political ecology has now flourished into a multifaceted and diverse field of study. Figure 3.3 demonstrates this by laying out the key interrelated elements through which

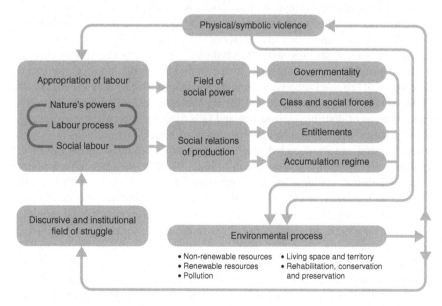

Figure 3.3 The framework of political ecology.
    Source: Adapted from Watts and Peet (2004).

political ecology perspectives seek to understand and explain the uneven degradation of both environments and human lives.

Whilst work in political ecology has a deeply embedded concern for inequality and injustice, and also has engaged closely with grassroots activism around, for example, mining and oil and other resource extraction, it has taken some time for connections to be drawn with environmental justice frames. However, as environmental justice has become more global and more diverse in its scope, we have increasingly seen a productive tying together of the process-based framework of political ecology with the more overtly normatively orientated and politicised world of environmental justice. As we saw in Chapter 2, environmental justice framing has moved globally 'southwards' into the predominant geographical regions of political ecology analysis, whilst political ecology work has moved globally 'northwards' (Keil 2003), extending its scope to include socio-ecological relations in wealthy countries and in urban and industrial settings, the key domains of established environmental justice activity. Smith (2006: xiv) nicely captures this complementarity:

> when complemented by an environmental justice politics, which is less internationally focused and less theoretical but more politically activist in inspiration, political ecology becomes a potent weapon for comprehending produced natures.

Where this coming together is most evident is in the emerging sub-field of urban political ecology (Heynen *et al.* 2006; Keil 2003). This has applied the theoretical resources of political ecology, particularly those rooted in a Marxist understanding of nature–society relations (drawing on the work of David Harvey), to urban settings with the ambition of 're-naturing' urban theory. Whilst sharing with environmental justice frames a central focus on the social unevenness of urban environmental quality and access to resources, Heynen *et al.* (2006) argue that the environmental justice literature is limited by its focus on liberal notions of distributional justice and fails to grasp how urban environmental inequalities 'are integral to the functioning of a capitalist political-economic system ... and are produced through the particular capitalist forms of social organizations of nature's metabolism' (ibid.: 9). The theoretical framework of urban political ecology has been presented as a 'manifesto' with ten key elements (summarised in Box 3.4). Perhaps key amongst this set of concepts and propositions is that of power. Social power relations are seen as having multiple forms, not only economic and political (as in the frameworks discussed above) but also material (the power of things and objects) and discursive (the power of words and images). In combination these constitute 'power geometries'. Material social relations are seen as operating in and through the transformations and metabolisations of the 'natural' environment which support and maintain urban life (water and food as well as computers and buildings). The environmental transformations taking place in cities are seen as part of 'power struggles' around class, gender or ethnicity rather than independent of them, with

Box 3.4 The urban political ecology manifesto

1 Environmental and social changes co-determine each other. Environments are combined socio-physical constructions that are actively and historically produced, both in terms of social content and physical-environmental qualities.
2 There is nothing a-priori unnatural about produced environments such as cities. The urban world is a cyborg world, part natural/part social, part technical/part cultural, but with no clear boundaries, centres or margins.
3 The type and character of physical and environmental change, and the resulting environmental conditions, are not independent from the specific historical, social, cultural, political or economic conditions and the institutions that accompany them.
4 All socio-spatial processes are invariably also predicated upon the circulation and metabolism of physical, chemical or biological components.
5 Socio-environmental metabolisms produce a series of both enabling and disabling social and environmental conditions. While environmental (social and physical) qualities may be enhanced in some places and for some humans and non-humans, they often lead to a deterioration of social, physical and/or ecological conditions and qualities elsewhere.
6 Processes of metabolic change are never socially or ecologically neutral. The urbanisation process reveals the inherently contradictory nature of the process of metabolic circulatory change and the inevitable conflicts that infuse socio-environmental change.
7 Social power relations are particularly important. It is these power geometries, the human and non-human actors, and the social–natural networks carrying them that ultimately decide who will have access to or control over and who will be excluded from access to or control over resources or other components of the environment.
8 Questions of socio-environmental sustainability are fundamentally political questions.
9 It is important to unravel the nature of the social relationships that unfold between individuals and social groups and how these, in turn, are mediated by and structured through processes of ecological change. In other words, environmental transformation is not independent from class, gender, ethnic or other power struggles.
10 Socio-ecological 'sustainability' can only be achieved by means of a democratically controlled and organised process of socio-environmental (re)construction. The political programme of political ecology is to enhance the democratic content of socio-environmental construction by identifying the strategies through which a more equitable distribution of social power and more inclusive mode of the production of nature can be achieved.

Source: Summarised from Heynen *et al.* 2006: 11–13.

patterns of empowerment in turn reflecting inequalities in urban environmental conditions:

> These metabolisms produce socio-environmental conditions that are both enabling, for powerful individuals and groups, and disabling, for marginalized individuals and groups. These processes precisely produce positions of empowerment and disempowerment.
>
> (Heynen *et al.* 2006: 10)

This power-laden production of material environments is also wrapped up with the mobilisation of particular discourses and understandings which seek to define what is natural, normal, abnormal, factual, rational and irrational. Power is embedded within dominant knowledge systems and the ways in which different modes of evidence and reasoning are valued and devalued – connecting back to earlier discussion of the politics surrounding alternative evidence claims.

Making sense of process

These various theoretically oriented frameworks, constituting varieties of claim about process, leave us with much to comprehend and consider. Other relevant perspectives could also have been included, from the utilitarian explanations of neoclassical economics (see Chapter 4 on waste), through core-periphery theory, to the heterogeneous human/non-human actors/actants of actor-network theory (see applications of this approach in Bickerstaff and Agyeman 2010; Holifield 2009). Looking across the explanatory frameworks we have considered though, some commonalities can be found:

- environmental inequalities are not explained simply, but are produced through multiple interwoven processes;
- there is a need to understand history and the operation of processes over time rather than just in the here and now;
- local processes leading to environmental inequalities are part of, and inter-related with, wider processes operating at regional through to global scales;
- different forms of inequality and injustice interact.

Beyond these points we can find different emphases – between culture and identity, economic and political structures, materiality and discourse – and different degrees of abstraction and universalisation. Writing in urban political ecology tends to universalise (Holifield 2009), seeing patterns of environmental inequality as always and everywhere serving to disadvantage those who are less powerful and economically, culturally or politically marginalised. The political-economy frameworks of Faber and Pellow generalise in similar ways, but within a tighter focus on the US corporate and political context and in relation to patterns of pollution, waste and risk. To what degree such process explanations *can* readily travel – between, for example, the socio-environmental domains covered in this book or across global geographies – remains, however, an open question. The conclusions of Chapter 2 might suggest that they cannot without some degree of re-contextualisation.

Summary

This chapter has covered much ground in examining three key elements of environmental justice claim-making: first, claim-making concerned with notions of justice – normative judgements about how things ought to be; second, claim-making about evidence – descriptions of inequalities in how things are; third, claim-making about process – explanations for why things are how they are, or how inequalities and injustice are produced. This simple organising scheme hides much complexity within it. Justice can be defined in terms of patterns of distribution, participation and recognition, each distinct but also interconnected. And each also requires specification as to the terms in which it is to be applied – which population and geography, which dimensions, which criteria or principle. Evidence of inequality can take both quantitative-statistical and qualitative-experiential forms, in which methodological complexities are inherent and politicised processes of evidence production and evaluation are embroiled. Explanations of process can be found in different theoretical frameworks, with different emphases and inclusions, but each recognising the multiple interacting processes involved, their historical embedding and operation across scales.

The value in having covered all of this territory is that we now have a rich set of resources with which we can: (i) begin to make analytical sense of particular cases of claim-making and how they are put together; (ii) develop our own perspectives on particular environmental justice concerns and contexts; and (iii) formulate critiques of the perspectives, evidence and arguments of others. In the rest of the book we will find much to bring these resources to bear on, and at the beginning of each of the following chapters I have included a list of particular elements of the framework that are to be emphasised in considering the substantive environmental justice concerns of that chapter. We will start in Chapter 4 with a focus on waste and siting, which will most directly work with and explicate the claim-making framework that this chapter has introduced.

Further reading

David Schlosberg provides the clearest and most persuasive account of the justice in environmental justice, in particular in his book *Defining Environmental Justice* (2007). Chapters 12–14 of David Harvey's text *Justice, Nature and the Geography of Difference* (1996) are also packed with insight and telling analysis. The discussion of evidence raised some fundamental questions about knowledge, science and expertise; reading work by Andy Jamison, Sheila Jasanoff, Alan Irwin or Brian Wynne would help fill this out. More practical matters of quantitative analysis of environmental inequalities are comprehensively covered by Liu (2001). A special issue of the journal *Environmental Justice* (2010, vol. 3, no. 4) focuses on participation and access to the legal system.

For further reading on questions of process, go to the key authors cited – Bullard, Pulido, Schnaiberg, Pellow, Faber. For following up on the rather involved ideas in urban political ecology there are excellent special issues of *Urban Studies* (2003, vol. 8, no. 4) and *Antipode* (2003, vol. 35, no. 5), as well as the edited collection of Heynen *et al.* (2006). Various process-oriented perspectives are also examined in the special issue of *Antipode* on 'Spaces of Environmental Justice' (2009, vol. 41, no. 4), which is also available as a book (Holifield *et al.* 2010).

4 Locating waste

Siting and the politics of dumping

Having spent some time in the previous chapters talking about concepts, theories, framing and frameworks, it is now time to get embroiled in more of the substance and detail of what environmental justice is all about. There is no better place to begin than with waste. Waste dumping was at issue in the protests that kick-started the US environmental justice movement and early accusations of environmental racism. The geographical patterning of hazardous waste sites was the focus of the very first environmental justice statistical studies that have since mushroomed across hundreds of papers, reports, theses and dissertations. And as we shall see, in many places around the world, and for many different communities, what happens to waste, where it goes, how it is handled, how it is processed, how it is contained – and the consequences of all this for health, safety and environmental quality – are deeply problematic and controversial. Waste is ubiquitously and continuously produced. We are all to some degree involved in divestment, in 'throwing things away' and in using products and services that generate unwanted stuff, some of which is particularly toxic and hazardous. Yet some people, places, and communities end up living with the back-end, the trash, the waste flows of modern society far more closely than others. Whether this is fair or just and, if not, what should be done about it therefore constitute core concerns for environmental justice analysis.

Waste provides us with many opportunities to explore the analytical themes introduced across Chapters 1–3. Too many in fact, so in structuring this chapter I have been selective, emphasising some themes over others, as outlined in Box 4.1. There is also inevitably selection involved in which cases or examples and which parts of the world are examined. As with other chapters, the research base and established activist framing of the US features strongly, but I have also deliberately sought to draw from waste controversies in other parts of the world, including where transnational flows of waste have been at issue. To ease us into thinking about why and how justice figures in waste controversies, we begin with three particular iconic cases from three different parts of the world.

Box 4.1 Waste and analytical themes

By discussing waste and siting issues at different scales we will be able to explore:

- the three core forms of justice – distribution, procedure and recognition – and how these are deployed in practice in waste disputes;
- the evidence, justice and process elements of environmental justice and how these can be aligned together in different ways in making claims about patterns of waste facility siting;
- the contested and sometimes overtly political character of evidence production and evaluation;
- the tensions that can emerge when making judgements about the injustice of a situation and therefore the objectives to be sought after; for example, when communities appear to actively welcome rather than resist waste facilities;
- the international scope of environmental justice concerns and how these relate to inequalities in regulatory and management standards between nations.

Resisting waste: three cases

There are many documented disputes about waste siting. Sometimes these have been positioned within an environmental justice frame, sometimes not. In each case, though, local resistance is generally associated with a decision process – to establish a new waste site at a particular location, expand an existing site, modify its operating conditions or undertake some other form of change. This provides a distinct, spatially focused decision event and, it follows, a political opportunity to make challenges and potentially to have influence on the future geography of waste and risk in that locality. The three specific cases we are about to examine all take this form, each being concerned with decisions about hazardous wastes and the development of a waste disposal facility. I have selected these cases because in each country – the US, Scotland, Taiwan – they have become emblematic of environmental justice protest – iconic cases that have each demonstrated the political saliency of an environmental justice frame. Each case is discussed before making comparisons between them.

Warren County: dumping in Dixie

In North Carolina in 1978 a haulage company illegally dumped 31,000 gallons of PCB-contaminated oil along 270 miles of roads, doing so at night to avoid detection. When in the early 1980s this contamination was discovered, the state authorities decided to scrape up the roadside soil and transport it in 600 truck-loads to a new hazardous waste landfill in Shocco Township, Warren County. Once this contaminated soil disposal was completed, it was planned that the landfill would remain open to accept commercial hazardous industrial wastes (Sasz and Meuser 1997). Warren County was one of the poorest counties (97th out of 100) in one of the poorest states in the US, with 65 per cent of the population being African-American, rising to 75 per cent in Shocco Township (Shrader-Frechette 2002). The people of Warren County objected. This was not their waste that they had responsibility for. There would be few benefits from hosting such a facility, but instead many expected and feared risks to health and safety. Why should a poor, predominantly black community be selected to take these risks, and the stigma of being associated with hazardous waste?

Protest actions were started up in the face of a complete lack of consultation or participation by the local community in the decision to choose Shocco Township as the location of the waste dump. Local African-American church groups were particularly active in developing a programme of non-violent direct action (such as protest marches and sit-ins), drawing on the traditions of the civil rights movement and bringing in activists from across the country (Bryant and Hockman 2005). The protesters distrusted the assurances of state officials that the site would be safe. The Environmental Protection Agency was allowing PCB wastes to be dumped only seven feet above the water table, a much lower separation than normally required. In one action 414 protesters were arrested as they attempted to prevent the passage of trucks to the site. The case became a cause célèbre, defining the start of the environmental justice movement in the US and claims that environmentally significant decisions could have racist intent. The protest did not entirely succeed; the waste ended up in Warren County, but a far wider political process that continues to the present day was set in train.

Greengairs: dumping in Scotland

Greengairs is a small village in North Lanarkshire, central Scotland, near three other villages with a combined population of about 1,500 people. By the early 1990s these communities had become surrounded by eight waste landfills in what were once large opencast quarries. Among these eight sites was the biggest land-fill in Europe, a massive hole in the ground for dumping waste. For many years local people had lived with the noise, smells, infestation and leaching from this intense concentration of waste, transported from urban communities across a wide region of Scotland (Dunion 2003). In 1996 the community's recurrent objections to the operation and impacts of these sites turned into a full-scale protest against a proposed site extension. In a distinct echo of Warren County, this protest was catalysed by a plan to dispose of polychlorinated biphenyl (PCB) contaminated soil coming from elsewhere. Indeed, the toxic waste was not only coming from England, and therefore crossing, as a result of recent political devolution, an increasingly significant political border, but from as far away as Hertfordshire, one of the wealthiest 'home counties' near London. The waste was being transported to Scotland because disposing of it in England had become problematic due to a tightening up of the Environment Agency's rules on accept-able PCB concentrations. In Scotland the rules were weaker, providing an opportunity for cheaper and easier disposal of the waste.

The local community at Greengairs was working class, an ex-mining commu-nity, not racially distinct nor the very poorest, but through making alliances with the national group Friends of the Earth Scotland (FoES) 'Greengairs' became the first high-profile environmental justice case in the UK. Local people complained that they already had enough dumps, that it was deeply unfair to keep on adding more and they had put up with more than could be reasonably expected. FoES argued that 'these are a group of communities who have to endure some of the worst examples of environmental injustice' (Indymedia 2004). The local commu-nity's previous attempts to negotiate with site operators to minimise local impacts had had little impact, as had representations at public inquiries into extensions of landfill operations in the area, each of which was considered case-by-case rather than in their accumulation. They were outraged that the managing director of the waste disposal company was also on the board of the Scottish Environmental Protection Agency (SEPA), which regulated and inspected the landfill operations. There had been no consultation locally about the proposal to take the PCB waste, so instead local people initiated their own engagements, blocking the road to the site, waving placards, stopping lorries and getting local and national media cover-age as a result, much of which picked up on the environmental injustice theme. The protests had an impact. The company commissioned an independent report that made 28 recommendations for the improvement of both landfill operations at the site and relationships with local people, and also advised against the move-ment of contaminated soil from England. All the recommendations were accepted by the company. SEPA also decided that the PCB concentrations in wastes allowed at the site should be reduced, which meant that the soil could not be dumped there in any case.

Orchid Island: dumping in Taiwan

Taiwan had been for many years a military dictatorship, under which political activism of many forms was suppressed. In 1987, when military rule finished and democratic institutions were established, civil rights and environmental move-ments flourished. These two streams of activism came together around the case of nuclear waste on Orchid Island, also known as Lanyu (Huang and Hwang 2009). Orchid Island is located off the south-east coast of Taiwan and is the home of the Yami, an indigenous aboriginal people numbering about 3,000. In the late 1970s an interim nuclear waste repository was set up on the island by Tai-Power, a state organisation, with the expectation that it would be feasible in due course to dump the waste in a deep ocean trench near the island. Dumping at sea was a form of nuclear waste disposal then practised by other countries, but it was subsequently banned (Fan 2006a). Even so, the waste continued to be stored on the island, and concerns grew about rusting containers, leaks and contamination and the risks these posed to health and the environment. Mobilising around this issue, activists complained that the Yami had never been consulted about the waste site and never been involved in its planning; in fact, they had been bullied and deceived (Fan 2006b). The government was accused of pretending at the time of construction that the facility was a fish canning factory rather than a nuclear waste store. Protesters argued that Orchid Island had been chosen because it was hidden away and home to a small powerless indigenous group. The Yami were taking all of the risks of nuclear waste on behalf of the rest of Taiwan, without even getting the benefit of the electricity generated, which was consumed only on the mainland. Attempts by Tai-power to compensate the Yami were rejected as being no replacement for the damage done to their culture and the spiritual values they gave to their landscape and environment. Discourses of environmental injustice and claims of deliberate environmental racism were used, picking up, in part, on the language that was by then well established in the US (Huang and Hwang 2009).

The government increasingly began to accept these arguments, issuing a statement in 2003 that explicitly referred to the violation of environmental justice principles. Having agreed to remove the waste from Orchid Island, the search for an alternative destination was initially international in scope – North Korea, China, Russia and the US all being considered. This prompted some furious reactions, particularly to the prospect of nuclear waste being sent to North Korea, a pariah state in terms of its nuclear weapons development. Finding a site within

Figure 4.1 The nuclear waste facility on the south side of Orchid Island, Taiwan.
    Source: Chris Stowers.

the territory of Taiwan therefore became necessary. In the latest developments the government has passed a law requiring any siting to be approved by a county-wide referendum, and for incentives to be provided to host communities. Two aboriginal townships have reacted positively to becoming potential sites, but intense debates continue as to the legitimacy of the decision processes being used.

Comparing the cases

What can we draw out of these three cases of protest against waste decision-making and siting from different places and parts of the world? There are similarities and differences between them. In each case, risks from particularly hazardous wastes were at issue; there was distrust of the reassurances about safety from authorities and company experts; the protesters deployed all the techniques of local political mobilisation against financially and politically more powerful actors; and both local people and activists working at regional and national levels became involved in making their voices heard. In these ways we can say that each case is typical of many forms of protest action that erupt where environmental risks and locally unwanted land uses are at issue. However, what is distinctive is that an environmental justice frame specifically became mobilised in the arguments and discourses of the protest groups – in each case for the first time within their national context. Through this framing each dispute became not only an issue of risk and land use but also a question of justice and fairness, and these words became a key part of the claims being made.

If we examine the forms of justice being argued about in each of these waste disputes, we also find similarities. In Chapter 3 it was explained how the justice in environmental justice can take different forms – distributional justice, procedural justice and justice as recognition – and how in practice all three forms of justice can be intertwined together in the arguments of activist groups. Accordingly, and as outlined in Table 4.1, we can use these categories to identify how, within the case-study contexts, different justice arguments were articulated.

There are clearly claims about *distributive injustice* in each case. These are expressed geographically in terms of one place and community having to take an unfair burden on behalf of others, and receiving few benefits in return, but also in racial and cultural terms in the case of the particular population profiles of Warren County and Orchid Island. Claims of *procedural injustice* are articulated in terms of lack of community consultation and participation, the failures of decision making processes and restricted access to information. *Justice in recognition* is part of each case, but central particularly to Orchid Island, with the indigenous Yami seeing their status denigrated and their distinct value system and right to self-determination disrespected by the siting of the radioactive waste dump. The justice claims are multidimensional, not resting on a single notion of justice but rather forming, as Schlosberg (2007) argues, a closely integrated combination.

Table 4.1 Forms of injustice articulated in each of the three cases

Case	Distributive	Procedural	Recognition
Warren County	Location in a poor, black community. Risks and impacts locally concentrated, but few benefits.	Lack of consultation and local involvement in decision to take contaminated soil.	Racist targeting of location and assumption that poor, black community would not object.
Greengairs	Gross over-concentration of landfills in one place. Accumulated risks and many dis-benefits for a working-class community.	Lack of consultation and local involvement in decision to take contaminated soil. Failure of decision making processes to account for accumulated impacts. Lack of impartiality by regulator.	Lack of respect for an already degraded and stigmatised place.
Orchid Island	Location of all Taiwan's nuclear waste in a remote, indigenous community. Local risks and impacts and no local benefits.	Lack of consultation and participation of local people. Secrecy and deliberate deceit about nature of site.	Yami seen as peripheral to the Taiwanese mainstream, a weak and backward indigenous group. Distinct cultural and spiritual values not respected.

Whilst there are these shared characteristics across the three cases, we can also see how the resistance against waste siting decisions is very much situated in *geographic, historical and political context*. As discussed in Chapter 2, the environmental justice frame has been applied and adapted flexibly in different parts of the world, and we can see differences between the southern US, Scottish and Taiwanese contexts that have been important in shaping each experience. The history and end of military dictatorship in Taiwan, the historic struggle for civil rights in the US and the devolution of Scottish government are part of both the dynamics of resistance that emerged in each case and the arguments about justice that were articulated. The local histories of how the specific communities perceived their past treatment by state officials and/or private companies were also important, emphasising that each protest was not just about a specific decision event, but also what came before and prefigured it.

Through these cases we can also draw out something of the *multi-scalar nature of environmental justice protest*. Geographers such as Williams (1999), Towers (2000), Bickerstaff and Ageyman (2010) and Kurtz (2002) have developed a powerful line of analysis which sees scale as central to the way that environmental justice politics is practised. Local disputes about pollution and risk do not stay simply as local disputes but are actively interpreted and represented as part of a broader social pattern – of discrimination, of cultural disrespect or exclusion from decision-making – which operates not just locally but over wider spatial scales. To be successful, they argue, activists need to use scale strategically, for example, by making the specific case relevant to the scales at which decision-making power is exercised and therefore political opportunities are to be found. We can see this multi-scalar quality in the Orchid Island case, in particular. The arguments and claims about nuclear waste disposal ranged right across the geographic scale, from the local distinctiveness, identity and rights of the Yami, through to national Taiwanese energy policy and international relations and responsibilities for dealing with nuclear waste (Huang and Hwang 2009). In the Greengairs and Warren County cases there are also ways in which the local disputes were actively interpreted as regional and national problems, symptomatic and illustrative of wider patterns rather than being unique. For Dunion (2003), for example, Greengairs is just one stop on a 'Dirty Scotland Trail' where an industrial legacy of pollution and contamination has unjustly blighted the lives and landscapes of many places and communities across Scotland. And as we shall examine in the next section, following the Warren County conflict specific local accusations of environmental racism also became part of a wider critique of systematic patterns of inequality concerning where waste sites were located, not just in North Carolina but across the whole of the US.

Unequal patterns of waste site locations

In kick-starting the definition and mobilisation of environmental justice activism in the US, Warren County also catalysed a stream of analysis that, as discussed in Chapter 3, has become central to much environmental justice scholarship. This quantitative analytical work looks for general patterns in the locations of environmental features relative to patterns in socio-demographic data. Coming out of Warren County, this initially took the specific form of looking for general patterns in the locations of waste sites in relation to racial profiles. The particular proposal to dump hazardous waste in a predominantly African-American community raised the question of whether or not this was part of a wider systematic pattern. Were such waste sites distributed across different types of communities without any distinctive racial bias? Or was there distinctive concentration and racially structured inequality at work?

Answering this question in the US context has proved to be particularly controversial and charged, and, partly for that reason, we will examine the progression of the empirical work in some detail in this section. The debate that evolved through the 1980s and 1990s around the methods, outcomes and implications of this work also provides a way of illustrating the close connections between the different elements of environmental justice claim-making discussed in Chapter 3, in turn demonstrating the ways in which different resolutions of what constitutes justice and injustice, and claims of evidence, can be arrived at. Having outlined the different phases of empirical analysis, we will then move on to some discussion along these lines.

First-wave studies

The first statistical studies of waste site locations in the US emerged directly out of the Warren County controversy and the debate this case initiated. Two pieces of research were particularly influential: the reports produced by the US General Accounting Office and the United Church of Christ – see Box 4.2.

These two 'classic' reports (Sasz and Meuser 1997), claiming striking racial disparities, are constant reference points in the US environmental justice literature and provided the foundation for a surge of other studies through the 1980s

---

Box 4.2 The two early studies of waste facility locations in the US

US General Accounting Office (GAO) (1983): Commissioned by Congressman Walter Fauntroy, who marched with the Warren County protesters, this research covered the eight south-eastern states. It found that three of the four hazardous waste landfills located within these states (including the Warren County site) had host communities that were majority black, and that at least 26 per cent of the host population had incomes below the poverty level. This socio-demographic profile was significantly different from that of the wider comparison population.

United Church of Christ (UCC) (1987): This was the first study covering the whole of the US, examining the locations of 415 commercial hazardous waste facilities. It compared ZIP code (postcode) areas with and without waste facilities and found that areas with waste facilities had double the percentage of minority populations when compared to areas without waste facilities. Predominantly black and Hispanic communities also hosted three of the five largest facilities in the US. Race was concluded to be 'the most significant among variables tested in association with the location of commercial hazardous waste facilities', forming a 'consistent national pattern' (ibid.: xiii).

---

and early 1990s that examined site location patterns in different ways for different parts of the US – see Sasz and Meuser (1997), Goldman (1994, 1996), Liu (2001) and Noonan (2008) for tabulated listings and reviews. These are categorised by Williams (1999) as the 'first wave' of empirical research, focused on revealing the extent of inequalities across the US and being concerned primarily with 'outcome-orientated' or distributive injustice, in which evidence of unequal patterns was the primary concern.

During this period the growing body of research did not remain on library shelves separate from the groundswell of environmental justice politics, but rather co-evolved closely with it. Most striking was how highly politicised claims of *environmental racism* emerged, fed by the statistical evidence of biased siting patterns at national and regional scales and articulated within local struggles against particular waste sites and decision-making processes. These claims were captured most starkly by Bob Bullard's highly charged phrase 'dumping in Dixie' (Bullard 1990), which equated the ways in which waste was being dealt with in the US with the many years of struggle against slavery and racism in the American South. The environment was slotted into the history of campaigning against racist and discriminatory politics in the US, to make claims that there were systematic discriminatory practices that were deliberately and intentionally locating waste and other polluting sites in poor black and minority communities. The argument went that, on repeatedly encountering 'not in my back yard' (NIMBY) resistance in white middle-class areas – empowered, environmentally aware, well resourced and well connected – waste companies and authorities were taking the 'easier route' of deliberately 'putting it in blacks' back yard' (the PIBBY syndrome; Bullard 1990), where resistance was assumed to be less likely and less effective.

Whilst these claims of racism were rooted in a deep foundation of critique of the practices of the state and powerful business organisations – and everyday experience of discrimination in housing, health and employment – specific evidence of racist intent in waste siting practices proved hard to come by. A widely quoted consultants' report to the California Waste Management Board, which advised on the communities least likely to resist the siting of waste-incineration facilities, provided some support for the deliberate targeting thesis (Cerrell Associates, Inc., 1984). This report advised that in the California context the ideal population for hosting a waste incinerator would be rural, with a workforce engaged in heavy industry, on low incomes, with limited formal education, and whose 'native tongue was not English'. Apart from that, though, evidence of discriminatory intent was characterised by Heiman (1996: 405) as 'largely circumstantial', and early attempts to establish otherwise through legal challenges did not prove successful.

Second-wave studies

These questions of causation, and whether or not race was the primary factor involved, rather than class, became central to a 'second wave' of statistical studies. Williams (1999) sees these as replacing the 'outcome-orientation' of the first wave with a greater 'process-orientation' – although this was more of a supplement than complete replacement. The second-wave research questioned the foundation of evidence on which environmental racism accusations were based, and challenged various aspects of the influential early studies. Criticisms were made of the methods used, the choice of spatial units, the neglect of other demographic variables, the use of comparison populations and statistical tests – which, as discussed in Chapter 3 and spelt out in detail by Liu (2001), are all open to judgement and selection. Repeat analyses using the same data as the early studies (or updated versions) were completed, and as a result of making alternative methodological choices they produced different results. For example, Anderton *et al.* (1994) repeated the national study of the UCC, using different geographic units and different modes of comparison to establish whether or not the population profiles near to commercial hazardous waste sites were distinctly different from those of other areas. Their conclusion was markedly different from the UCC's: 'no consistent national level association exists between the location of commercial hazardous waste TSDFs and the percentage of either minority of disadvantaged populations' (ibid.: 232).

The authors of the UCC study, and others (Goldman 1996; Mohai 1995), in turn hit back at the Anderton *et al.* study, pointing out that it was funded by WMX Technologies, the largest commercial handler of solid and toxic wastes in the world (and by implication could therefore not be trusted to be impartial), and mounting trenchant counter-challenges to the alternative methodological choices that had been made. Goldman and Fitton (1994) completed another repeat of the UCC study in 1994 using up-to-date data and concluded that the concentration of minorities in host communities had not disappeared but had in fact increased from 25 per cent in 1980 to nearly 31 per cent by 1993. A further repeat in 2007, 20 years after the original study, which, it is argued, used the most sophisticated methodologies of all, also found even stronger disparities emerging over time – see Box 4.3.

Box 4.3 Toxic Wastes and Race at Twenty (Bullard *et al.* 2007)

This repeat analysis employed 1990 and 2000 census data and distance-based methods to investigate the extent of racial and socio-economic disparities in the location of commercial hazardous waste sites across the US. Disparities are examined by region and state, and separate analyses are conducted for metropolitan areas, where most hazardous waste facilities are located. At a national level, using 1990 data, the analysis showed that the percentage concentration of people of colour is highest within 1km of facilities (47.7 per cent), decreasing as the analysis takes in a wider area (see Figure 4.2). Analysis using 2000 data found a still higher concentration, with the percentage of people of colour within 3km rising from 46 to 56 over the period. It found that for 2000 more than 9 million people lived within 3km of the 413 facilities across the US. Of these, more than 5.1 million were people of colour. Neighbourhoods with facilities clustered close together have higher percentages of people of colour than those without site clusters, and also higher levels of poverty. Forty of the 44 states and 105 of the 149 metropolitan areas with hazardous waste facilities have disproportionately high percentages of people of colour within 3km.

The study concludes that race continues to be a significant predictor of hazardous waste facility locations when socio-economic factors are taken into account, and that 'although the current assessment uses newer methods that better match where people and hazardous waste facilities are located, the conclusions are very much the same as they were in 1987' (p. xi).

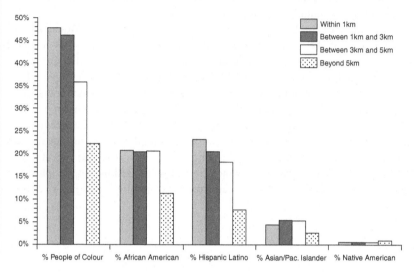

Figure 4.2 Percentage of people of colour living near to hazardous waste facilities in the US (1990 census data).
Source: Adapted from Bullard *et al.* (2007).

The second wave of studies in this way set off a process that has continued ever since in which a greater awareness of methodological limitations has, on the one hand, given plenty of scope for critique by those who want to be sceptical or challenge findings that they do not like and, on the other, led to the use of ever more sophisticated methods of quantitative analysis in the effort to establish more definitive outcomes (see further discussion of similar methodological developments in Chapter 5). In addition though, the question of temporal sequencing in causation was raised by the second-wave studies specifically as a challenge to the environmental racism claims of discriminatory intent. This came to be known as the 'chicken or egg' question.

Chicken or egg?

The chicken or egg question is about what came first. Was it the case that when waste facilities were first sited the host communities were poor and minority dominated? Or did they become so *after* the siting decision had been taken? This was presented as a crucial test of discriminatory intent. If at the time of siting the host communities were not predominantly made up of a minority population, it was argued, then racism accusations could have no foundation. A study by Been (1994) was the first to take this question on in quantitative terms, although it encountered many methodological difficulties. She revisited the GAO study, and another by Bullard (1983) for the City of Houston, and compared the socio-economic status of communities hosting hazardous waste sites when these were first established. Her analysis confirmed that the site locations *were* poor and minority-dominated at the time of siting, but introduced a market dynamics hypothesis to explain changes since this date. This hypothesis was that, once in place, a hazardous waste facility would lower land and property values, discourage upper-income households from living in the area, and make the area more attractive to poorer residents (who might also 'happen' to be from racial minorities). Tracking changes from 1970 to 1990, she found some evidence to back up this hypothesis for the Houston area – levels of poverty and the number of black residents went up and property values went down – but not for the four sites covered by the GAO study.

Oakes *et al.* (1996) undertook a national study in 1996 along similar lines, finding no evidence to back up claims of systematic bias in the siting of commercial hazardous waste sites (as in the Anderton *et al.* 1994 study conducted by the same team) and, in their longitudinal analysis, that changes in the host communities were no different from changes in populations in other industrial areas. In other words, the presence of the site did not seem to have a particular effect on population dynamics. Been and Gupta (1997) conducted a further national longitudinal study and reached slightly different conclusions. They found that there *was* a racial effect in siting, although one which skewed distributions towards Hispanic rather than African-American populations, but no income-related effect towards the poorest populations. They could also find no evidence for the market dynamics hypothesis operating after siting.

Regional and city-scale studies have continued to examine the temporality of waste site location and population patterns in different parts of the US, producing varied results. Four examples of research papers are outlined in Table 4.2, each examining different places and different categories of waste facility. Just across these four examples we can see results that show waste sites concentrated in black or Hispanic areas when these were first established, and/or an increasing concentration on these populations over time, but also some very different patterns and dynamics at work.

The complexities involved in making sense of such variability in findings provide a clear example of the general points made in chapters 1 and 3 about how evidence of environmental inequalities cannot simply be taken at face value. A set of correlation or regression coefficients may superficially look compelling and convincing, but they have the status of claims rather than truths, constructed out of methodological and analytical choices. In the face of conflicting results, systematically comparing and evaluating how these were arrived at therefore

Table 4.2 Examples of longitudinal studies of waste site locations in the US and their key findings

Study	Area	Types of waste sites	Key findings
Stretesky and Hogan (1998)	Florida	Superfund sites	Concentration in black and Hispanic areas increased over time from 1970 to 1990
Baden and Coursey (2002)	Chicago	Hazardous and solid waste disposal sites	No evidence of concentration in black areas, historically or currently; growth of Hispanic population over time in areas with waste sites as part of general trend
Hurley (1997)	Metropolitan region of St Louis, Missouri	Abandoned hazardous waste sites	Current pattern of racial concentration largely reflects demographic change; at time of siting very few of waste sites were in majority African-American locations
Pastor *et al.* (2001)	Los Angeles County	Toxic storage and disposal facilities	Initial disproportionate siting in racial minority areas matters more than disproportionate minority move-in over time; areas with sites are also low income, with few home owners, and show racial 'churning' over time

makes much sense, for example, in the form of a meta analysis (see Ringquist [2005] for an example), as does trying to understand whether differences in findings are an artifice of methodological choices or a reflection of different spatially and historically constituted underlying conditions and processes (Noonan 2008). But as some of the arguments between researchers coming at these questions either from an activist background or with the backing of industry demonstrate, in public debate evidence claims become a resource, mobilised for or against a preferred position. In other words, such evidence contests are not just about who has the 'better' science but also about politics and power. As Sasz and Meuser (1997: 113) observe in reviewing the development of empirical environmental justice research: 'political agendas not only motivated, they subsequently determined researchers' horizon – what questions would be asked, and how they would be asked would be largely determined by American race politics'.

Environmental racism or markets? Analysing positions

One of the key ways in which evidence of inequality becomes a resource to be interpreted through an ethical or political lens is in its integration into justice arguments. In Chapter 3 an analytical framework was outlined which identified three interconnected elements of environmental justice claim-making – claims about evidence, about concepts of justice and about process. The debates about patterns of inequality in waste facilities that we have worked through demonstrate how alternative integrations of evidential, justice and process arguments can be put together. Table 4.3 uses this framework to characterise three different positions on the question of racism and the injustice of waste facility siting. These positions are rather simplified but draw on broad distinctions in the relevant literature and associated discourse.

The first position, *intentional environmental racism*, characterises the early claims of environmental justice activists – or rather how these claims came to be understood and represented – that there was a racial bias in siting patterns and deliberate racism at work in siting decisions. This position has a view of distributive justice which sees the deliberate burdening of one racial group more than another as fundamentally wrong. Whilst politically powerful, we have seen how this position became rather boxed in, or 'bogged down' (Heiman 1996: 406), by a lack of direct evidence of racist intent, and by the statistical complexities of trying to resolve the chicken or egg question. Given that the question of 'what

Table 4.3 Three different positions on environmental racism and waste facility siting and their evidence, justice and process components

Position	Evidence of inequality (how things are)	Justice (how things ought to be)	Process (why things are how they are)
Intentional environmental racism	There is a systematic bias towards waste sites being located in host communities with a high percentage of racial minority population.	Sites should be distributed more equally, without bias towards racial minorities.	There is deliberate racism within siting decision-making processes, targeting communities that are less powerful and politically organised.
Institutional environmental racism	There is a systematic bias towards waste sites being located in host communities with a high percentage of racial minority population.	Sites should be distributed more equally, without bias towards racial minorities.	There is institutional racism embedded within housing markets, planning processes and the cultures of state and private organisations.
Market dynamics	There may or may not be differential racial patterns in the location of waste sites. If there are, these reflect the association between race and class.	Whatever is determined by the operation of land and housing markets, and economically efficient siting decisions.	Waste companies and residents make decisions that reflect their needs and interests within the working of land and housing market processes.

was just' came to rest on whether or not intentional racism was at work, if this could not be proven then the possibility of environmental injustice seemingly went away – as did the prospect of successful legal challenges to siting decisions on these grounds.

The second position, *institutional environmental racism*, takes a different stance in not letting the determination of racism or injustice rest on proving deliberate intent. This has a similar view of the evidence and of justice to position one but sees racial bias in siting patterns as being the consequence of wider structural processes of institutionalised racism. The patterns of bias are fundamentally wrong and unjust, it is argued, regardless of how they have been produced. Pulido (1996) has most forcefully adopted this position. She argues that trying to prove intent through quantitative analysis is far too limiting, and that racism has to be understood instead as integral to collective social life in the US, part of the way in which private and public organisations work, infusing culture, politics and economic structures. As she asks:

> isn't the housing market skewed by racial discrimination as dozens of studies have shown? … is it not racist that African-Americans, Chicanos/Latinos are disproportionately represented in the ranks of the poor and therefore all the more vulnerable to pollution through depressed land values, their role as low-wage workers and limited political power?
>
> (ibid.: 149)

In this way, Pulido is critical of the first position and its focus on intentional racism in specific decision-making practices, as this limits the scope of what is seen to be racist, making it something abnormal rather than ideological and endemic to the way in which US society operates. Others have made similar or related points. For example, Heiman (1996) argues that waste siting outcomes are a function not just of racial but also of class oppression; Williams (1999) and Pellow (2007) argue that particularising environmental injustice as purely a matter of local decision-making disconnects it from the fundamentally uneven and unjust multi-scale social processes that capitalism sustains (as discussed in the 'process' section of Chapter 3).

The third position, *market dynamics*, which characterises the perspective of environmental racism sceptics (Jeffreys 1994), (some) industrialists, right-wing politicians and economists, takes a very different evaluative stance. Here siting outcomes are explained as a consequence of the decision-making of rational actors operating within a market system. A utilitarian view of justice is taken, in which the rational actions of individuals, firms and public officials lead to optimal and fair outcomes. Evidence of bias in patterns is to a large degree inconsequential, as inequalities of various forms are an inevitable (and unproblematic) outcome of the differential resources, needs and interests that these actors have. Firms make economically efficient decisions to locate where land is cheap and where operating costs are low; households make decisions about where to live and where they can afford. As Noonan (2008: 1156) notes: 'when an economist sees a landfill surrounded by poorer residents, he or she is as likely to see households shrewdly saving on rent, as to see households suffering environmental injustice'. Inequalities in proximity to waste sites, if they exist, originate from the 'natural' functioning of supply and demand and the impersonal working and logic of market forces. Jeffreys (1994: 682–3) expresses this very clearly in arguing that areas that are now condemned as evidence of environmental racism should rather be seen as places where 'minorities were given their first access to the American Dream ... Even with the pollution and the low-wage jobs, their lives were greatly improved.' He continues: 'how ironic that the very economic forces that eventually spawned the civil rights movements would be condemned as environmental racism today'.

So we have here three ways in which claims about evidence, justice and process can align to arrive at some fundamentally different perspectives on waste facility siting. Their integration is not random, but rather is evidently a matter of ideology or worldview. If you have a worldview that is fundamentally uncritical of American capitalism and the uneven way that market processes work out, then position three immediately resonates and appeals. If you see contemporary American (and global) capitalism as fundamentally problematic in the uneven consequences it produces across racial and class groups, then position two is immediately persuasive. There are nuances and sophistications that this discussion hasn't been able to convey, but ultimately there are such fault lines in how environmental justice claim-making is evaluated, and these fault lines cannot simply be resolved by more sophisticated empirical research, or by getting people with competing perspectives around a debating table.

As a final comment it is relevant again to point out the distinctiveness of the US environmental justice frame and its focus on race. Quantitative studies of waste site locations have been undertaken in other countries, but, for all of the contextual reasons outlined in Chapter 2, these have not necessarily focused on the race question. For example, in Box 4.4 the outcomes of studies undertaken in the UK are outlined. Again, varying patterns have been identified, but in each case the empirical work has focused on poverty or deprivation. Here

---

Box 4.4 Quantitative analysis of waste site locations in the UK

Over the last five years a series of studies have examined the distribution of waste sites in the UK, for different areas, for different types of sites and using different methods. Varied results have, partly as a consequence, been produced. In all cases the focus has been on deprivation or poverty as a social variable.

Scotland: Fairburn *et al.* (2005) analyse the locations of all 224 waste landfill sites in Scotland, finding no relationship with deprivation due largely to their more rural location. Richardson *et al.* (2010), in a more sophisticated study of municipal landfills, do find a relationship with deprivation. They also, uniquely amongst UK studies, investigate the temporal dynamics of siting, concluding that deprivation was evident when sites were first established but has also become more concentrated subsequently.

Wales: Walker *et al.* (2008) examine the distribution of various types of waste site, finding a highly differentiated geography. The urban locations of recycling and waste transfer sites mean that they have high numbers and proportions of deprived people living near to them. Landfills are located further away from population concentrations and are not biased towards the deprived. Higgs and Langford (2009) examine just landfill sites in Wales, using various population estimation methods, finding that deprived populations are not found very close to landfills but are at a moderate distance away – a 'halo effect'.

England: Damery *et al.* (2008a) examine various types of waste site in the north-west of England, concluding that more deprived populations are more likely to be living nearer to waste sites than the less deprived, except in the case of landfill sites where it is the *least* deprived populations who are more likely to live nearby. Fairburn *et al.* (2009) find a strong clustering of active waste sites in deprived areas in South Yorkshire. Walker *et al.* (2005) find a strong bias in the locations of waste incinerators in England towards deprived populations.

inequality in these terms appears to matter, in contrast to the US where for some observers (of the position three variety in particular) demonstrating an inequality in waste site location by income or class is inconsequential (Sasz and Meuser 1997).

Displacement, toxic imperialism and environmental blackmail

In the three cases of siting controversy outlined at the beginning of this chapter, we could see processes of displacement at work. Unwanted and risky wastes were to be moved from their source and relocated at a distance from where the waste was produced. We can see this displacement in simple cartographic terms, but also in a political or cultural sense. The distancing is away from heartlands or cores into places that are politically and culturally marginal. In two of these cases the attempted displacement was international in scope – contaminated soil to be moved from England to Scotland, nuclear wastes from Taiwan to, at one point, North Korea – emphasising the extended scales at which waste flows and waste 'solutions' can now operate. We saw in the last section that explaining why displacement operates in the way that it does, and produces particular outcomes, can become complicated. Whilst we explored competing claims about racism and the working of land and housing markets, there are further dimensions to be considered.

Hazard export and the 'race to the bottom'

There has been an extended debate since the 1970s about the extent to which the tightening of environmental standards in more wealthy countries has the consequence of pushing environmentally risky and polluting activities into places with lower standards and weaker regulatory regimes, inequalities in standards thereby promoting inequalities in waste burdens. A very limited example was seen in the Greengairs case, where the waste company was transporting contaminated soil to Scotland because the limits on acceptable contamination levels were, for a period, laxer there than in England. Roll this out to a bigger international scale and contrast the regulatory standards applying in, say, the US, Germany or the Netherlands with those applying, or very weakly enforced, in central Africa, central America, Eastern Europe or parts of Asia, and the thesis that companies might be persuaded to displace hazardous industries and hazardous wastes to less demanding regulatory regimes and cheaper places to do 'dirty business' becomes persuasive (Clapp 2001; Pellow 2007). This has variously been called 'toxic imperialism', the 'race to the bottom' (of environmental standards), 'pollution havens' and the 'export of hazard'. Low and Gleeson argue in this context that 'the global political system, composed of competing nation states, may be unfitted for the task of guaranteeing environmental justice' (Low and Gleeson 1998: 120), pointing to the ways in which trade liberalisation and economic globalisation have allowed greater discretion in locating production activities and disposal of wastes, and trade agreements have generally failed to recognise environmental concerns as a reason for constraint.

The toxic imperialism thesis has been examined and debated at some length (within and outside of an environmental justice frame), and various attempts have been made to establish its existence and extent. Harvey (1996) and others (Newell 2005) find significance in the infamous leaked 1991 memo from Lawrence Summers, then Chief Economist of the World Bank, which began: 'Just between you and me, shouldn't the World Bank be encouraging more migration of dirty industries to Less Development Countries?', continuing later: 'I think the economic logic of dumping a load of toxic waste in the lowest-wage country is impeccable' and 'under-populated countries in Africa are vastly under-polluted' (quoted in Harvey 1996: 366). Although it has been argued that the memo was written as a deliberately provocative challenge to conventional neoclassical economic thinking in the World Bank, it nonetheless clearly expresses the types of logic that can lead to the movement of wastes and risky industries from one part of the world to another.

Attempts to gather more material evidence of these processes have taken the form of empirical studies of macro patterns, and the documentation of particular cases of damaging waste flows by researchers and activist groups. Macro-scale economic analyses have largely focused on the movement of industrial activities and trade flows and have produced mixed and contradictory results. For example, Ederington *et al.* (2005) and Spatareanu (2007) both find evidence of significant (but relatively small) effects from environmental regulations on trade and foreign direct investment patterns. In contrast, Wheeler (2000) contends that the 'race to the bottom' model is not supported by empirical research and that its basic assumptions are flawed. Communities in developing countries, he argues, will over time act to control pollution effectively, and well-informed consumers and investors will give value to environmental performance even in developing countries, providing an incentive for multinational companies to maintain globally high standards.

Wheeler's assumptions – based on narrow modes of analysis (Clapp 2002) – are not borne out by well-documented examples of the ways in which hazardous wastes have found their way into 'host' countries and had damaging impacts on nearby people and environments. Many of the worst cases go back to the 1980s, when there was a large growth in the international movement and trade of toxic wastes into poorer countries. The publicity and outrage generated by environmental NGOs at this time helped to bring the international Basel Convention into force in 1989. This, through embodying a key justice principle (Okereke 2006), made dealing with hazardous wastes the responsibility of the producer country, banning international trade between rich and poor countries under most defined circumstances (Clapp 2001). Whilst significantly limiting the scale of the problem, the Basel Convention has by no means eradicated it. Not all countries are signatories and various loopholes have been exploited, including 'sham recycling' where wastes are fraudulently labelled as for recycling rather than disposal. Illegal movement undoubtedly goes on and sometimes this is detected and prosecuted, as in the highly publicised Trafigura case, see Box 4.5.

Box 4.5 Trafigura and toxic waste dumping in Africa

In 2006, London-based oil traders working for the company Trafigura illegally arranged for the disposal of a shipload of hazardous wastes in the Ivory Coast in Africa. The dumping took place after the company had failed to get rid of the waste cheaply in the Netherlands by describing it as routine slops from tank-cleaning. Trafigura pumped the toxic waste back onto the tanker, which sailed to the Ivory Coast where the black slurry was dumped in landfills around Abidjan. Serious medical consequences followed for local people, with reports that 15 deaths, 69 hospitalisations and more than 100,000 medical consultations were linked to the dumping.

The company vociferously denied these claims and took legal action to try to block media reporting. In 2009, after much international outrage and the revealing of email exchanges which confirmed that there was intent to make a profit by disposing of the waste at low cost, Trafigura was forced to pay compensation totalling £30 million to thousands of Africans who needed medical treatment. The emails revealed that the traders hoped to make profits of $7 million a time by buying up what they called 'bloody cheap' cargoes of sulphur-contaminated Mexican gasoline, processing the fuel and disposing of the resulting waste at low cost. The payments settled a civil legal action brought by London lawyers and followed another payment of £100 million to the Ivorian government to clean up the waste. In 2010, the company was also fined £840,000 for the initial concealment of the dangerous nature of the waste in the Netherlands.

See http://joana-morais.blogspot.com/2010/05/trafigura-accused-of-bribing-witnesses.html for a more detailed account, including copies of email exchanges.

Pellow (2006, 2007) provides detailed and extended accounts of a range of such cases exposed by the work of environmental activists, often across transnational networks linking producer and destination countries. Along with others working in an environmental justice frame, he has particularly focused on the movement of electronic wastes, which are exported in vast quantities to developing countries on the grounds that they are to be productively recycled or re-used (Iles 2004). The Basel Action Network (2002, 2005) has traced the end-destinations of e-wastes in Asia and in Africa, documenting examples of the dangerous reprocessing of wastes by women and young children, dumping in informal landfills and the burning of wastes, both of which can lead to air and water pollution from the lead, cadmium and mercury and other toxic materials in electronic components. However, obtaining anything like a reliable picture of what e-wastes are being traded and where they end up is extremely difficult (Robinson 2009). Lepawsky and McNabb (2010) provide a recent review of the field and attempt to map the global flows of e-waste in the face of considerable methodological problems. They conclude that the picture is geographically very complex, with multiple types of movement, stages of reprocessing and reuse of retrieved materials. They do find some support in their data for the pollution havens thesis, with an overarching trend towards e-waste flows ending up in low-income countries in Asia, but they also note problems in conceiving of e-waste as an end-point in a linear chain of production, consumption and disposal:

the existence of trade and traffic networks for e-waste means that, by definition, some form(s) of value exist or are created after disposal takes place.

Even as waste, the harmful health and environmental effects of e-waste pro-
cessing are neither evenly distributed nor do they remain contained in a single
locale as the notion of a pollution haven might suggest.

(ibid.: 191)

Waste as a commodity and community consent: saying yes?

Wrapped up in talking about waste in terms of trade and value is the complicating
fact that waste is a form of commodity. What is unwanted by one person or
organisation may be actively valued by another, either because it can be productively processed, recycled or re-used, or because the operation of a waste facility
can generate jobs or other forms of income that bring benefit. This is tricky territory for environmental justice advocacy, as it can confound easy assumptions that
waste flows and waste sites will always be seen as something bad within a
community, rather than maybe as something good. Various examples have been
explored in the literature. For example, Pellow (2002) analyses the case of a
waste incinerator in Robbins, Illinois, where a long-standing all African-
American municipal government was faced with very divergent local viewpoints
on the prospect of raising significant revenue from an activity perceived, by
some, to be a serious threat to health. Lake (1995) explores the conflict between
applying universalised environmental justice principles 'and the equally compelling principles of local autonomy and determination' through two cases: that of a
proposed major landfill site in West Virginia and that involving the siting of a
paper recycling plant in the Bronx, New York. Bickerstaff and Agyeman (2010)
analyse the case of a 'ghost ship' full of toxic materials that was routed to the
north-east of England for scrapping. This was framed by Friends of the Earth and
other campaigners as an environmental injustice of toxic dumping in an economically deprived and already polluted community, and by local politicians and
trade unionists as a significant economic opportunity to dismantle a ship under
tight regulatory conditions that are not applied in other parts of the world.

Such situations have been analysed in various ways. Bullard (1992) calls them
cases of 'environmental blackmail', where communities are so desperate for jobs
and income that they are prepared to trade their environmental quality and health
at almost any cost. Harvey (1996), quoting his own earlier work, notes the
'intriguing paradox' of how incentives to trade-off or give up environmental
quality are differentiated between rich and poor: 'the rich are unlikely to give up
their local amenity at any price, whereas the poor, who are least able to sustain
the loss, are likely to sacrifice it for a trifling sum' (Harvey 1973: 81). What, more
analytically speaking, these cases open up is a distinction (and often a tension)
between outcome and process in justice terms. If a low-income minority or indigenous community, through an inclusive and open democratic process, decides
that it wants to host a waste facility, then should our evaluation of the outcome
shift? In other words, in making judgements about a pattern of inequality does
procedure matter as much as, or maybe more than, the distributional outcome?
When earlier we examined the debates around empirical studies of waste site

location in the US, it was clear that deliberate racist targeting of waste site locations was wrong, and, for some, the workings of the land and property market could be found equally unfair. But how about the cases where it appears that community consent to host a waste facility is happily and fairly given? Does criticism then have to be adjusted or suspended?

In the Orchid Island case outlined at the beginning of this chapter, the ongoing story of the attempts to find a location for the disposal of nuclear waste in Taiwan demonstrated exactly these questions. The government had belatedly tried to make decision-making processes more environmentally just by passing legislation requiring a regional referendum to be held before a new site location could be approved, also giving communities incentives to become involved as candidate sites. Two aboriginal townships responded positively, but amongst much controversy as to the legitimacy of their actions and whether decisions were being taken on proper democratic grounds. Such approaches are being used in various parts of the world to try to solve waste siting problems. For example, in the UK a process of community 'volunteering' for a radioactive waste repository is currently under way (Committee on Radioactive Waste Management 2007) – and typically it is remote, peripheral and marginal places and communities that are becoming involved as potential hosts (Blowers 1999). In terms of the legitimacy of the process, much rests on exactly how democratic it is seen to be (Petts 2005; Watson and Bulkeley 2005). How is the 'community' defined? How inclusive, open and transparent are participatory opportunities? What resources are available? And so on. These are all the key questions of procedural justice discussed in Chapter 3, and spelt out specifically in relation to hazardous waste siting processes by Hunold and Young (1998). When forms of monetary incentive or community benefit provision are involved, these can also be interpreted very differently: as reasonable and appropriate forms of compensation for bearing the disbenefits of waste disposal on behalf of wider society, or as an attempt to pay off opposition through a barefaced bribe.

Given all these distinctions, it is not surprising that processes generating 'yes in my backyard' responses have been internally divisive. In Canada (Stanley 2009) and the USA (Gowda and Easterling 2000; Ishiyama 2003; Shrader-Frechette 2002), cases of waste disposal within Native-American communities have been particularly shot through with tensions and contradictions. In the 1990s the US government embarked on a volunteering process to find a location for a major nuclear waste store, in theory opening up their search to any state or community in the country (Hoffman 2001). Most immediately rejected the possibility, but 20 responded more positively, 16 of which were Native-American tribal governments. Through a narrowing down process the Mescalero Apache tribe became the first to sign up for the waste siting programme. A positive vote by the tribal council was then overturned in a wider referendum, only for that to be overturned again by a second referendum in favour of continuing negotiations with the nuclear utilities. Hoffman (2001: 469) recounts how this and similar processes in other tribal areas have been deeply divisive, 'bringing into sharp relief the difficulties in establishing truly representative procedures for meaningful participation in decision-making processes'. In the case of the proposed siting of a nuclear waste repository in Skull Valley, Endres (2009) contrasts the argument that the Goshutes tribal government had exercised their sovereign right to host the facility with arguments from opponents within the Goshute community that they had not been properly consulted, that there was corruption amongst the tribal leadership and that any financial benefits from being involved in the process had only gone to the elite in favour of the development. This means, Endres argues, that claims of environmental injustice cannot be dismissed in this case, particularly given the long history of colonialism and 'corporate oppression' that had left the Goshutes already vulnerable and economically dependent on forms of toxic development.

But context matters, and, in reviewing a range of potential environmental justice cases on tribal lands, Krakoff (2002) argues for a nuanced analysis rather than generalisations which see environmental injustice operating for all environmental decisions in all tribal areas. Some cases stand up to analysis much less well than others, she argues, such as those where tribal authorities have applied tougher environmental standards than normal and have developed regulatory regimes that are sensitive to their cultural and religious traditions. Strengthening tribal peoples' abilities to address their own environmental problems and overcome their historic and economic deprivation, she concludes, is more important than finding a 'single evil actor' to blame, or acting externally to impose solutions from the outside.

From redistribution to prevention

Through this chapter what perhaps at first appeared as a straightforward focus for environmental mobilisation has become steadily more involved. Applying justice concepts and claims to the siting and management of waste has thrown up an increasing number of tensions and dilemmas in how siting processes and outcomes should be evaluated. We could go further still, for example, by bringing questions of intergenerational justice into decision-making on radioactive waste, as waste facilities sited now will have consequences for hundreds if not thousands of years into the future (Shrader-Frechette 2002). Whilst this emphasises the challenges involved in making sense of what environmental justice means in practice, it also points to another response to the problem of locating waste – to put it simply, don't produce the stuff in the first place. For many environmental justice activists and observers there is a fundamental problem in trying to distribute wastes somehow more fairly, one which, as we have seen, can lead to waste flows being displaced into ever more peripheral, economically marginal and politically weak communities. As Pellow comments, 'the environmental and environmental justice movements in the North have unwittingly contributed (at least partially) to the flow of destructive multinational corporate operations and hazardous wastes to the South' (Pellow 2007: 14). This presents the real danger of a continual ratchet effect, in which resistance to waste siting, as it spreads through successful environmental justice activism, continually pushes the waste 'somewhere else'.

The answer has therefore to be to direct attention not just towards particular facility siting decisions, but more fundamentally towards the processes producing waste in the first place. Consequently, waste minimisation, clean production, clean energy generation and alternative green modes of economic activity are all themes that have increasingly fallen within an environmental justice framing, often linked to a sustainability agenda. At the weaker end of the spectrum this takes the form of arguments for increased reuse and recycling rather than incineration or landfill dumping, particularly of domestic and commercial wastes. For example, Watson and Bulkeley (2005) analyse the justice principles embedded within recent waste management policy in the UK and conclude that the waste hierarchy (which favours reduction, reuse and recovery over disposal) and the following of principles of proximity and self-sufficiency when disposal is required effectively enact concepts of both intra- and intergenerational environmental justice. As they point out though, the crucial test is how effectively the implementation of these principles is carried through into decision-making at a local level.

At the more radical end, agendas go much further. Heiman (1996: 411), for example, makes a strong call for 'democratic control of the means of pollution' through people gaining knowledge and self-help skills so 'that they can take control of the forces affecting their own lives'. Pulido (1994) argues for extending notions of procedural justice upstream into the domains of normally 'private' production decisions of corporate bodies if universal global rights to a clean environment are to be aimed for. Faber (2008: 252) calls for 'productive justice' rather than 'distributive justice', in which 'not-in-my-backyard' is transformed into 'not in anyone's backyard'. For Lake (1995) focusing on self-determination, as the 'only guarantor of environmental justice', is the necessary route out of the dilemma of communities facing stark choices between jobs and environmental degradation. By 'democratic participation in the capital investment decisions through which environmental burdens are produced and communities are affected' (ibid.: 170) siting decisions, he argues, become neither necessary nor desirable, and 'turf-based battles' over the distribution of waste and pollution are eliminated.

These are compelling positions, but ones which have to be continually tempered by the pragmatics of dealing with waste that exists in the here and now. For example, highly dangerous nuclear wastes exist and their management is a problem that cannot be wished away, even if we might argue vociferously that no more should be produced and therefore work against the construction of new generations of nuclear power stations (Walker 2009b). Searching for just and fair waste siting processes and distributive outcomes is therefore still necessary, even if at the same time challenges are made to the institutional structures through which waste production is normalised and sustained.

Summary

In this chapter we have examined waste and siting issues both as a key focus for environmental justice research and activism and as a way of working with the analytical framework introduced in Chapter 3. Through focusing on a range of debates about waste, hopefully it has become clearer why it is useful to make distinctions between different forms of justice concept, why it is necessary to see evidence as a form of claim-making, and why understanding process and causation is also part of the construction of justice claims. We have worked through an important phase in the evolution of the environmental justice debate in the US, a period during which quantitative analyses of patterns of socio-spatial inequality became high profile, controversial and problematic. We have seen how such statistical evidence became interpreted through different political lenses and integrated into conflicting understandings of how patterns of racial inequality in waste siting have been produced – through racist intent, institutionalised racism, or the operation of market dynamics.

In tracing waste flows and waste controversies in different parts of the world, we have clearly seen the global relevance of environmental justice framing (picking up on the themes of Chapter 2), and the ways in which problems and disputes in different places are interconnected. Processes of displacement can lead to wastes being moved to geographically, politically and/or culturally distant locations, incentivised by weaker regulation and lower costs, although the scale and extent of so-called 'toxic imperialism' or 'hazard export' processes are disputed. We saw that a key, but problematic, justice issue concerns the conditions under which destination communities take the waste flows of others. If open and informed consent appears to be given, for example, to obtain the economic gains of waste as a commodity, does this mean that justice is satisfied? Does procedure in such instances matter more than outcome? Such tensions are in theory resolved if attention is focused on waste prevention and reduction, rather than on its distribution and displacement, but long-term goals of prevention have to be reconciled with the continued existence and production of wastes in the present and the foreseeable future. Waste therefore emerges from this discussion as a key and problematic environmental justice issue, and as meriting careful reasoning about the rights and wrongs that are involved.

Further reading

There are some excellent introductions to waste as a phenomenon, a generic problem and a feature of modern society, material that this chapter did not cover; see Thompson (1979) on 'rubbish theory' and O'Brien (2008). The three cases outlined at the beginning of the chapter have been detailed at much greater length. For Warren County see Bullard (1990) amongst many others. For Greengairs and other Scottish cases see Dunion (2003) for an excellent account from an activist's perspective. Environmental justice in Taiwan is reviewed by Huang and Hwang (2009), whilst qualitative research on the Orchid Island case can be found in Fan (2006a, 2006b). For other US case studies see the detailed accounts by Pellow (2002) of 'garbage wars' in Chicago, and also activist accounts in Bullard *et al.* (2007). For an interesting analysis of an Irish incinerator dispute that examines why justice discourses were not deployed see Davies (2006), and for a successful mobilisation against an incinerator proposal in the UK see Dodds and Hopwood (2006).

The controversy over empirical studies in the US is covered in detail by many others. Liu (2001) is a great source for those interested in methods and more technical issues. Goldman (1996) provides an entertaining discussion of the politics of competing evidence claims, and Pulido (1996) a fundamental critique of how racism was being conceptualised. Both are in an excellent special issue of *Antipode* (1996, vol. 28, no. 2). Gender issues and waste have not been discussed in this chapter but provide the focus for Buckingham *et al.* (2005).

For the international movement of hazardous wastes see books by Pellow (2007) and Clapp (2001), including discussion of the work of transnational activist networks, and also a special issue of *Environmental Politics* (2009, vol. 18, no. 6). On the complex issues involved in reconciling environmental justice and processes of volunteering and community consent, both Shrader-Frechette (2002) and Lake (1995) do a more sophisticated job than I have been able to.

5   Breathing unequally

Air quality and inequality

Breathing good quality air is a fundamental requirement for a healthy life. Around the world pollution from industry, traffic and other sources diminishes the quality of breathed air but not in a consistent way, and we might expect some systematically uneven social patterns to exist. Air pollution might well 'follow the poor' (Beck 1998), but does it always do so, and if so why? How about other demographic, social or cultural groups, how might we make an appraisal of their exposures to poor quality air? And what constitutes a just situation in terms of how air quality is socially distributed, scientifically appraised and managed? These provide core questions for this chapter, which deals, as the last chapter did, with a topic readily positioned within an environmental justice frame. Air quality has been the focus of the work of many environmental justice groups in the US. Examples include multiple campaigns in New York on childhood asthma and its relation to traffic, industrial and incinerator emissions (Sze 2007) and in the San Francisco Bay area on the health impacts of diesel emissions, pollution from power plants and oil refinery flaring (Pastor *et al.* 2007). In South Africa air pollution has been central to the environmental justice activism focused on emissions from controversial oil-refining operations in Durban (Barnett and Scott 2007), and in the UK to environmental justice research and campaigns initiated by Friends of the Earth over inequalities in emissions from industrial facilities and from waste incineration.

   Air quality enables us to undertake a wide-ranging exploration of environmental justice claim-making. The specific cross-cutting themes that I have chosen to emphasise in this chapter are outlined in Box 5.1, taking us from methodological questions through to the application of justice concepts. Because of the complexities involved, we will begin the discussion with a scheme that lays out the multi-dimensionality involved, working briefly through the elements of the framework introduced in Chapter 3. This will provide something of an overview and a guide to the more detailed discussion in the rest of the chapter.

Air quality and multidimensional claim-making

In Chapter 3 a simple framework of the key elements of environmental justice claim-making was laid out. This consisted of claims about evidence

Figure 5.1 Environmental justice campaigning on air quality.
Source: West Harlem Environmental Action.

Box 5.1 Air quality and cross-cutting themes

By discussing air quality as an environmental justice issue we will be able
to explore:

- the methodological complexities and challenges involved in deriving
  evidence of inequalities, in terms of both exposure to polluted air and
  the resulting impacts on health;
- the importance of understanding the social distribution of vulnerability
  to harm, as well as the distribution of exposure to poor air quality for
  different social groups;
- the contestation that can emerge around the scientific basis for
  evidence of patterns of pollution exposure and impacts;
- the historically constituted and geographically specific processes that
  have produced contemporary socio-spatial patterns in air quality;
- the interrelated distributional and procedural justice grounds on which
  claims of injustice are constructed by environmental justice activists.

(how things are) and process (why things are how they are), the linkage between them and what they imply for justice (how things should be). Within each of these broad categories various further 'sub-dimensions' were identified as more or less relevant to particular environmental justice concerns. For air quality the distinctions between distributional patterns of exposure, vulnerability and responsibility are particularly important. Air quality inequality is more than just a spatially structured problem of where polluted air is to be found and who lives in those locations. The impact of poor air quality on health and well-being is also a matter of the vulnerability of those breathing in the pollutants, and this vulnerability might well vary by age, gender, class, ethnicity and other socio-demographic characteristics. Also, given that air pollution is by definition produced by human actions,[1] questions of responsibility are similarly relevant. Pollution concentrations are not 'acts of god', and therefore establishing who is responsible, or who is to blame, is a pertinent dimension of justice claims.

In Table 5.1 the three dimensions of air quality distribution are set against the three elements of the claim-making framework – evidence, process and justice.

Table 5.1 An example of a set of claims made about air quality and its relation to poverty

	Exposure	*Vulnerability*	*Responsibility*
Evidence	The poor are most exposed to high levels of air pollution and breaches of air quality standards.	The poor are more vulnerable than others to air pollution impacts and therefore more likely to suffer ill health from breathing polluted air.	The poor are least responsible for the generation of air quality problems.
Process	Historically uneven patterns of land use change, the working of housing markets and an unequal ability to choose where you live explain patterns of exposure.	Poverty means people are less healthy generally, and less able to access good health care.	Low incomes mean less ownership and use of cars and a smaller contribution to air pollution.
Justice	Everyone should have the right to a minimum standard of clean air. Monitoring of compliance with standards should involve communities.	Standards should be set to protect the most vulnerable. Research processes should be participatory and community based. Risk assessments should incorporate the multiple dimensions of health problems.	Action should be taken and financed by those most responsible for causing air pollution. The polluters must pay.

1 Poor air quality *can* be produced by processes that are predominantly 'natural', such as dust storms or volcanic eruptions, but this does not constitute 'pollution' as such.

Within each cell of the table an example of a claim that fits with these interrelated dimensions is given, each orientated towards the relationship between poverty and air quality. These specific claims typify the statements or arguments made by environmental justice activists. Across the first row are various forms of evidence claim, each of which could be tested empirically. Across the second row are various forms of process claim, each in some way seeking to explain why inequalities related to poverty are to be found. In the third row are various forms of justice claim encompassing not just a desired pattern or standard of air quality distribution but also matters of procedural justice, involving demands for community inclusion in scientific appraisals and air quality management decision-making.

There is a lot to unpack in this table, and much which might be argued with or presented in alternative ways. It does, however, provide an initial opening up of the territory of this chapter and an opportunity to begin thinking through some of the challenges of generating evidence to support these claims, substantiating process explanations, or realising justice objectives. In the rest of the chapter all of these elements and their complexities will be discussed in more detail, beginning at the top left of the table with evidence of inequalities in exposure to air pollution.

Evidence of air quality inequality

Environmental justice research has examined patterns in the socio-spatial distribution of exposure to air pollution in many different ways to identify whether there are systematic and regressive biases at work. Studies have looked at patterns for countries, regions, cities and local neighbourhoods. They have focused on ethnic and racial groups, on indicators of poverty and deprivation, on children and older people. They have examined different pollutants and combinations of pollutants and linked their impacts to various potential health consequences. Most of the research that has been undertaken has involved the quantitative analysis of the social distribution of patterns either of emissions of pollutants from particular sources or of monitored levels of ambient air quality. Studies have related pollution data with data on the socio-demographic characteristics of geographically located populations to make claims of evidence, or otherwise, of degrees of inequality.

As discussed in general terms in Chapter 3, in embarking on an analysis of the social distribution of air quality there is a whole series of methodological choices to be made, extending from the types of pollutant and social data to be used to the scale and techniques of statistical analysis. Table 5.2 summarises some key choices and the complexities that can be involved, such as data not being available in the form needed and how to take account of the different impacts on health that different constituents of 'polluted air' can have.

The choices that are made in putting together and carrying out a study inevitably shape the scope and form of the evidence claims that can then be made and the knowledge that is generated – and, it follows, what knowledge is *s and complexities in the distributional analysis of air quality/pollution

oice	*Examples of choices*	*Complexities*
egories	Poverty, deprivation, ethnicity, race, children, older people	Data may not be collected or available, or may be out of date. Social data is usually generalised to particular irregular spatial units.
Pollutants	Nitrogen dioxide, Sulphur dioxide, particulates, ozone, carcinogens	Data may not be collected or available; the significance of potential health impacts varies; combining pollutant loads is problematic.
Type of data/ measurement	Ambient levels in the atmosphere; emissions from specific sources; average, peak levels, or exceedences of standards	Data may not be collected or available. Ambient data can be sparse, measured in different ways or based on models. Emissions data may be incomplete, unreliable or estimated.
Spatial scales	National, regional, city, neighbourhood	Data may not be collected at a particular scale; the density of available data may be low.
Temporal scales	Current patterns, past patterns, future potential patterns	Consistent comparable data may not be available on past emissions. Looking ahead involves modelling based on scenarios.

not generated. Choices are shaped by the nature of the frame that the work is being done within (for example, environmental justice instead of environmental health) and also by the institutional setting and resource context of the research. The complexities involved can limit the choices available and produce uncertainties in the outcomes of analysis.

At a broad level such considerations shape the global geography of the available evidence about air pollution inequalities. Not surprisingly, most is known about patterns in the US, where the politics of environmental justice have been most acute. There is also a body of work emerging in other developed countries, for example, in Canada (Handy 1977; Jerrett *et al.* 1997; Jerrett *et al.* 2001), New Zealand (Pearce *et al.* 2006), the Netherlands (Kruize *et al.* 2007) and Sweden (Chaix *et al.* 2006), reflecting the global movement of the environmental justice frame. In contrast, whilst it is known that levels of ambient air pollution can be far more severe in cities in the developing world, there has here been little systematic quantitative investigation of patterns of distributional inequality in such cities, for reasons that include severe limits on resources and the collection of data. We can note though that some of the inequalities of air pollution exposure in the developing world require little statistical evidence. Indoor pollution problems associated with the burning of polluting fuels on open fires or rudimentary stoves in the developing world are evidently focused on poorer communities, as they are structured by both reliance on cheap fuel sources and poor quality housing, with severe consequences for respiratory health (Dasgupta *et al.* 2006; Smith and Mehta 2003).

Focusing on the studies undertaken in developed country contexts, the very earliest analyses were undertaken in the US in the 1970s – interestingly predating the emergence of the environmental justice frame (Freeman 1972). With the arrival of environmental justice campaigning in the US, the choice of social categories centred resolutely on race, sometimes in combination with indicators of income or poverty, with the intent of evaluating the existence or otherwise of environmental racism – in a similar way to the studies on waste siting discussed in the last chapter. Only fairly recently has research in the US become socially more expansive, for example, focusing on children as a demographic sub-group through analysis of the location of schools in relation to patterns of air pollution (Pastor *et al.* 2006; Sexton *et al.* 2000). Varied geographic scales of analysis have been undertaken, from national studies looking across the whole country to regional or state level geographies and analyses for specific cities or metropolitan areas. While most studies have focused on current patterns, some have traced trends of change over time, using both past and current data (Korc 1996).

National level studies provide a broad picture of patterns of distribution across the national space but, given their spatial resolution, much local variation is inevitably hidden. In general, national studies in the US have concluded that racial minority populations are more concentrated in areas with higher pollution levels, but that poverty is less clearly and substantially a metric of inequality (Liu 2001). For example, Wernette and Nieves (1992) found that racial minorities were over-represented in areas designated by the EPA as not attaining designated 'criteria levels' of air quality, and that this over-representation was stronger than for indicators of poverty. These results therefore suggested that inequalities were not just the result of an association between minority status and low income, but that other racially structured processes were at work. Differences in the nature of the disparities between urban and rural areas have been highlighted. Liu (1998) found that the poor had a higher burden of poor air quality in both urban and rural areas, but that the bias towards racial minorities was predominantly in urban spaces. In contrast, Gelobter (1992), in a national study looking at particulate matter trends over time, found that in some years it was in fact the *highest* income groups that had the greatest exposure.

Regional and city level studies in the US have looked at air quality in many different ways – by using data on ambient levels of key pollutants, by focusing just on emissions from sites falling within the Toxic Release Inventory (a database of the most harmful emission sources across the country), by using different ways of aggregating populations and relating them to air quality or emission data, and by employing different statistical methods. Given this variation in study designs it is not surprising that a mix of resulting knowledge

Table 5.3 Examples of regional and city scale analysis of the social distribution of air pollution in the US

Citation	Place and scale	Pollutants	Key findings
Napton and Day (1992)	5 cities in Texas	Emissions of 7 criteria pollutants	Populations in high emission areas are more likely to be white, have more schooling and be home owners.
Bowen *et al.* (1995)	Ohio County and SE Cleveland	Toxic release inventory (TRI) data on releases to air	For Ohio County, strong associations between release amounts and racial minorities; for smaller scale units in Cleveland more differentiated. patterns, with correlation between releases and poverty rather than race.
Mennis (2005)	New Jersey	Air polluting facilities	Facilities tend to concentrate in high % minority neighbourhoods and enforcement is weaker in these areas.
Sheppard *et al.* (1999)	Minneapolis	Airborne TRI emissions	Higher % of poor people live near polluting sites; bias stronger for poor white than poor racial minority.
Pastor *et al.* (2004)	Los Angeles	Modelled concentrations of ambient air toxics and proximity to TRI sites	Children of colour are more likely to be exposed to health risks through school location; disparities in risks may be associated with diminished school performance.
Grineski *et al.* (2007)	Phoenix	Modelled criteria pollutants	Higher levels of air pollutants associated with lower neighbourhood socio-economic status, higher proportions of Latino immigrants, and higher proportions of renters.

claims have been produced. Table 5.3 summarises a selection of studies to demonstrate this variation in outcome and the contradictory conclusions that are drawn. High emissions are associated with both predominantly white *and* non-white communities. Sometimes poverty is concluded to be more important than race and sometimes vice versa, and results can shift as scales of analysis are altered.

Such variability leads Sheppard *et al.* (1999: 25), who review a range of studies examining patterns of toxic releases into the air, to conclude that results 'often conflict with one another' and that the 'very different results are at least in part a consequence of differences in data used and measure of potential exposure, applied to different kinds of places, at different geographic scales, with data of different levels of spatial resolution'. Any form of simple generalisation about what the body of analysis says about the relationship between air quality and social difference in the US is therefore problematic.

Looking to research in the UK, there is a similar diversity in the types and scales of study design, although there are also features that reflect the particular national setting. National studies have all focused on indicators of deprivation and poverty (and largely ignored race or ethnicity), reflecting the dominant wider framing of environmental justice in the UK discussed in Chapter 2 (Fairburn *et al.* 2005; McLeod *et al.* 2000; Mitchell and Dorling 2003; Walker *et al.* 2003; Wheeler 2004). Regional and city studies, for example, have similarly used a range of largely class or income related variables, considering different pollutants and sometimes constructing aggregate pollution measures (Brainard *et al.* 2002; Fairburn *et al.* 2009; Pye *et al.* 2001). Most studies show that air quality is worse for poorer or more deprived communities (Lucas *et al.* 2004), but this is not always and everywhere the case, or necessarily a simple linear relationship.

This can be demonstrated by looking more closely at some of these study results. Several of the national studies have found a U-shaped rather than a linear relationship between income/deprivation/poverty and air quality. For example, Figure 5.2 shows a distributional trend for England in which the very worst average nitrogen dioxide (NO_2) levels are in the most deprived population deciles (1 and 2). But rather than continuing on an improving trend, air quality worsens again for the *least* deprived population deciles (9 and 10), creating the (partial) U shape. Similar patterns have been found for other pollutants, for Scotland (Fairburn *et al.* 2005) and for Great Britain as a whole (Mitchell and Dorling 2003).

If we then look at the same data for Wales, a more striking 'anomaly' appears, in that, as shown in Figure 5.3, the lift upwards for the more wealthy (least deprived) population in decile 10 is so strong that they, rather than the most deprived, are found to live with the most polluted air in the country (a pattern found for other pollutants as well).

Such counter-intuitive results, which fail to comply with the assumption that pollution will be worst for the worst off, demand explanation, and it is to this that we now turn.

Explaining patterns of inequality

The evidence we have reviewed so far in some ways raises more questions than answers. Whilst some of the revealed variability in patterns of relationship between air quality and social difference might be explained by methodological

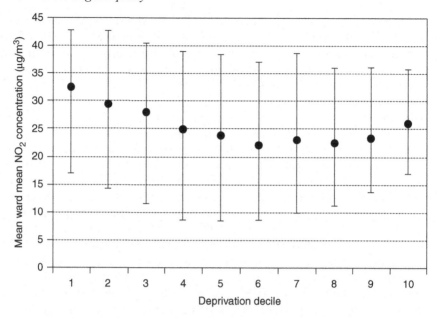

Figure 5.2 Distribution of nitrogen dioxide by deprivation in England, 2001.
Source: Walker *et al.* (2003).

Notes:
1 Bars denote 5–95 percentile range, N=8,414.
2 Each decile represents the average of ward mean NO$_2$, measured as an annual mean.
3 Deciles have equal total populations. Decile 1 is the most deprived, 10 the least
   deprived.

differences, it also suggests that we need to understand more about the underlying
processes through which these patterns have been produced. This takes us to the
second element of the framework laid out in Chapter 3 and in Table 5.1, that of
process and explanation. The task here is to interpret patterns of air pollution
distribution by understanding the interaction of pollutants and geographies in
their spatial and temporal contexts. Accordingly, if we take the U-shaped curves
found in Figures 5.2 and 5.3, these can be explained by the particular geography
of NO$_2$ as a pollutant, and by the historical evolution of socio-economically
segregated residential patterns in England and Wales. In England, concentrations
of deprivation are mainly found in urban and, particularly, urban-industrial
areas, a pattern which goes back to the growth of cities through periods of rapid
industrial development in the nineteenth and early twentieth centuries and the
emergence of a very uneven pattern of housing and environmental quality. Such
areas are also those with concentrations of transport and industrial emissions that
lead to elevated NO$_2$ levels. Rural areas in comparison are predominantly
wealthy, again an embedded historical pattern reflecting patterns of land ownership and the ability of those with resources to buy their way into the English

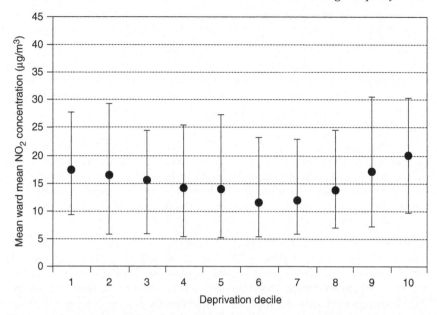

Figure 5.3 Distribution of nitrogen dioxide by deprivation in Wales, 2001.
       Source: Walker *et al.* (2003).

Notes:
1 Bars denote 5–95 percentile range, N=865.
2 Each decile represents the average of ward mean NO_2, measured as an annual mean.
3 Deciles have equal total populations. Decile 1 is the most deprived, 10 the least
   deprived.

'rural idyll'. They also have low levels of traffic and industry emissions and, given that NO_2 is a form of pollution that does not travel far or linger for long periods in rural areas, NO_2 concentrations are generally low outside of urban centres. This, at a simple level, explains why the very worst air quality in England (on the left-hand side of Figure 5.2) is associated with deprived populations. At a deeper level we could look to a framework such as that of urban political ecology (see Chapter 3) to trace how political and economic forces and structural differentials in 'power geometries' (Heynen *et al.* 2006) explain the uneven processes of urban and economic development.

However, urban areas are not universally deprived. Some of the most wealthy people in England have for a long time lived in the exclusive enclaves of its cities, for example, in the Kensington, Knightsbridge and Chelsea areas of London, and with urban regeneration and gentrification the retention of wealth in inner city areas has been actively pursued. This explains why the curve in Figure 5.2 turns upwards for the least deprived deciles on the right-hand side. Wealthy and empowered urban residents can ensure some elements of their quality of life – high quality housing, gardens, access to restaurants, theatres, galleries, etc. – but other aspects, such as exterior air quality, are less readily demarcated and controlled.

Moving to the Welsh situation, and the more extreme U-shaped curve of Figure 5.3, we have first to note that, in the light of comparatively low population and traffic densities, concentrations of pollutants are much lower than for England and exceedences of standards are very rare. In Wales the most deprived communities have traditionally not been in the major urban areas where most pollution is now generated, but in the South Wales valleys historically associated with coal mining along with some remoter rural areas. With virtually all coal mining shut down by the end of the 1980s, the study using data from 2001 showed that these valley communities are still deprived in relative terms but enjoy generally good air quality. The most polluted areas are now associated with traffic in cities, and the capital city of Cardiff in particular, and it is here where the wealthy middle classes of Wales have been drawn to live. Capital city employment opportunities, culture and quality of life act as attractors for the middle classes, whilst air quality is not so bad that it serves as an active repellent.

Studies in other places, at other scales and for other pollutants also demonstrate the need to explain socio-spatial patterns of air quality in spatial and temporal context. Kruize *et al.* (2007), for example, analyse the social patterning of exposure to nitrogen dioxide pollution around Amsterdam airport, finding a 'fairly equal distribution' across social classes, a situation they argue is the result of the historical interaction of market forces in the area and tight environmental regulation of local emission sources. Liu (1996) highlights the distinctive geography of ozone pollution, which, for various reasons and unlike all of the other major pollutants monitored and regulated, is found at higher levels in rural areas than in urban areas. For this reason, in the regions of New York and Philadelphia his study revealed that zones of higher ozone concentration were predominantly found in both white and higher income rural communities. In California, with its own distinctive social geography, Korc (1996) found that for ozone the most highly exposed group were Native Americans living outside of the urban core, followed by whites, Hispanics, Asians and African-Americans. Cutter and Solecki (1996), in a complex study of acute and chronic air pollution emissions in the south-eastern US measured at county level, particularly emphasise the need to understand the specificities of place-based processes. In testing the 'dumping in Dixie' hypothesis (see Chapter 4), they conclude that distributional patterns of emissions are varied and under certain conditions do disproportionately affect African-American communities. However, they also found high levels of emissions in other socio-economic settings (including some of the most wealthy counties) and identify a set of socio-spatial processes that explain the 'riskscape' of the region. These include 'regional urban-industrial growth, rural underdevelopment, the structure of industrial production (small facilities v large ones), proximity to transportation corridors and racism' (ibid.: 395).

This suggests that detailed historical studies are needed to understand patterns of urban, social and industrial change and their relationship to pollution over a long time scale. These are relatively rare but Hurley (1995) provides a fascinating example, tracing the history and social biases of environmental change in Gary, Indiana, from 1945 to 1980 (see Box 5.2). Pulido (2000) provides an equally

Box 5.2 Changing air pollution inequalities in Gary, Indiana

In the post-war era a major concentration of steel manufacturing meant that Gary ranked as one of the most polluted cities in the US. Its residential communities stretched south from the steel factories, with wind patterns determining which areas experienced the worst of the dumping of particulates and fumes. In the 1950s the areas closest to the factory, and the most polluted, were very mixed socially, reflecting different past waves of immigration and urban growth, although the wealthiest residents lived in areas to the south and east that were least affected by pollution. Over subsequent decades Hurley's data shows that particulate pollution became more skewed towards the poor and black community. By 1980 the black residents of Gary tended to live in the most polluted neighbourhoods, and a range of income indicators all correlated strongly and inversely with pollution levels.

All this is explained through a number of changes in the urban-industrial profile of the city. Whilst steel maintained its importance, changes in production technologies meant that pollution levels declined to some degree. However, other metal, chemical and power production industries moved into the city, producing new air, land and water pollution problems and new geographies of relation between pollution sources, dispersion patterns on prevailing winds and affected communities. During the same period Gary's population decentralised. Whilst the industrial centre moved westwards, population spread outwards towards the city fringes. The black population tripled in size, and racial conflict erupted as the mid-town area became densely packed and black populations moved into other areas of the city. The 'path of least resistance' where white incumbents were least effective in preventing racial mixing was towards the industrial areas of the north and west. A 'white exodus' into more distant cleaner areas to the south and east produced an increasingly differentiated socio-environmental geography.

Source: Summarised from Hurley (1995).

compelling analysis of the historical evolution of urban development and what she terms 'white privilege' in southern California. Such longitudinal analysis shows that whilst snap-shot analyses of current data reflect contemporary socio-spatial positions, these need to be understood as rooted in historically embedded processes.

Vulnerability and impacts on health: the 'triple jeopardy'

Most of the research that has been discussed up to this point has only taken evidence of inequality so far. It has analysed patterns of variation in air quality and linked these to patterns of variation in where people of different incomes, ethnicities and ages have their homes. Many such studies take it as read (and often only by implication) that consequences for health will then follow – that exposure to damaging air pollution will result and impacts on bodies will be felt. Whilst it might be reasonable to make this assumption, particularly where standards set to protect health are being exceeded, there are also reasons why evidence collection and analysis might seek to go further and to establish more explicitly the nature of unequal health consequences.

For those inclined to be sceptical of the significance of analyses of environmental inequalities (as we have seen in Chapters 3 and 4, some people and organisations undoubtedly are) there is much to be picked at in distributional studies of air quality and their reliance on simple geographies of proximity and coincidence as a proxy for unequal impacts on health. There is an evidential gap between being able to identify a spatial correlation between emissions/ambient levels and the aggregated characteristics of residential populations (as in the studies discussed above), and then assuming that these people have in fact breathed in significant amounts of polluted air and that this has had harmful effects. Emissions can be diluted and dispersed in complex ways that defy simple proximity measures; ambient measures depend on the density of monitoring points and can miss much local variation and peak concentration; people move around day to day and don't only breathe the air where they live; and much more besides. For such reasons, Buzzelli (2007) argues that 'probing the linkages' between unequal distributions and health consequences is the necessary next step towards a 'new framework' for environmental justice research. His review of 78 environmental justice studies between 1995 and 2003 found that none tested for health associations and only nine mentioned health implications at all, a deficit that he argues needs to be addressed.

Another way of seeing limitations in the simple geography of distributional studies is that they fail to focus satisfactorily on the interaction between differences in the air and differences in the spaces of the household and the body (see Walker 2009a for a broader discussion of different notions of space in environmental justice research). As argued in general terms in Chapter 3 and laid out in the second column of Table 5.1, vulnerability can be an important dimension in the construction of claims of inequality – and this is clearly the case for air quality, where campaigning groups have pointed to much differentiation in sensitivity to harm. They have highlighted that bodies that are young or old, that already have health problems or that are of particular ethnicities may be more vulnerable to breathing polluted air than others. Furthermore, people in households that are in poverty, that are struggling to maintain healthy lives in other ways and that have worse access to health care may experience more severe or longer-lasting compounded impacts from polluted air. The same level of pollution may, in other words, have different consequences for different people, meaning that inequalities in the socio-spatial distribution of pollution are not *all* that matters. Jerrett *et al.* (2001) refer to this as a potential 'triple jeopardy for health' in which poor socio-economic conditions interact with both poor health *and* a poor living environment, including, for some, poor air quality. For these reasons, a coming together of environmental justice analysis and air pollution epidemiology has been argued for (Lipfert 2004; O'Neill *et al.* 2003) and a growing number of studies have moved in this direction.

Environmental justice campaigning and research have increasingly interacted productively with expertise based in traditions of environmental or public health. Indeed, Faber (2008: 6) argues that environmental health activism was one of the founding branches of the environmental justice movement, and Coughlin (1996) has traced how the work of environmental justice activists has been influential in developing an increasing awareness amongst epidemiologists and other health professionals of the ethics and values that should guide their work. As he states, 'one example of the way in which the values of epidemiology and those of individuals and communities overlap is in the interrelation of epidemiology and the struggle for environmental justice worldwide' (ibid.: 67). Through the making of such connections health professionals in the US have become increasingly involved in undertaking studies on air pollution and inequality, in some cases working directly with minority and poor communities to produce data on environmental burdens and patterns of ill health (Dutcher *et al.* 2007; Loh *et al.* 2002). Similar alliances can be seen elsewhere, including, for example, in Germany where environmental justice has largely been interpreted as an issue of health inequalities (Hornberg and Pauli 2007).

One stream of work bringing environmental justice and public health expertise together has involved the development of more sophisticated, large scale data analysis, using increasingly involved statistical methodologies to attempt to isolate causal relationships between pollution exposure, ill health and social difference. Bard *et al.* (2007), for example, report on an involved study design examining the relationship between exposure to five air pollutants, two forms of health outcome and two measures of socio-economic status in the Strasbourg metropolitan area in France. Briggs *et al.* (2008), in a national scale study of England, focus in particular on the 'triple jeopardy' argument. They relate a general indicator of health status to a long list of environmental variables, including eight forms of air pollution emissions and ambient measures, reaching the guarded conclusion that 'whilst the triple jeopardy of deprivation, increased potential for exposures to environmental pollution and impaired health certainly exists ... the additive effects of deprivation and environment on general health status are usually not strong, and not always negative' (ibid.: 15). Grineski (2007) provides an example of a number of studies that focus on children. In a multivariate study of air pollution and the incidence of asthma in Phoenix, Arizona, she concludes that both air pollution (specifically ozone) and the quality of housing are important predictors of asthma incidence. This, she argues, may serve as a 'double jeopardy' such that 'poor asthmatic children in Phoenix often live in dilapidated rental homes located near freeways with mold, old carpet, and dusty yards' (ibid.: 369). Such studies demonstrate the complexities of the interaction between socio-demographics, living and environmental conditions and health status and the research challenges involved in realising the 'ideal union' (ibid.: 370) of environmental justice and epidemiological methods. As we shall see later though, whether pursuing ever more expert and involved study designs is the most appropriate route for addressing such complexities is open to debate.

The distribution of responsibility for air pollution

The third distributional dimension introduced earlier in Table 5.1 is responsibility. In Chapter 3 it was argued that, in addition to vulnerability, evidence related to patterns of responsibility can be included as a further combined element of environmental justice claim-making. Situations where there is a dislocation between those creating and those suffering from harm or disbenefit can be particularly significant in justice terms. For air pollution it has often been taken as read that this dislocation exists. Where pollution from industrial sources is at issue – either through local community campaigns against specific sites or in analytical studies using data on industrial pollution emissions – it is reasonably clear that the pollution source and its impacts are structurally disconnected. Responsibility for emissions rests with businesses, corporations and sometimes public utilities, whilst exposure is to adults and children living nearby. There can be arguments about the extent to which those (adults) living nearby might be employed in such industry, or benefiting economically in other ways. But there are far fewer 'company towns' or predictable patterns of workers living 'over the fence line' than in the past, and environmental justice campaigners have often stressed the degree of disconnection between company sites and nearby communities. There also can be arguments, as we have seen in Chapter 4, about the extent to which people living in more industrial and polluted areas have chosen to live there, deciding to accept worse environmental quality as part of their housing market behaviour (if they are part of the market, that is), so that pollution is not necessarily entirely 'imposed' on them. Even if accepted – and, as we have seen, there are many counterarguments – such considerations do not make that significant a difference to the gap between who is responsible for the production of industrial pollution and who is suffering the consequences of its 'consumption'.

For other sources of pollution though, this is not as clear cut. Pollution from traffic is the primary cause of poor air quality in many urban areas. Cars are driven by millions of people, including those who live in more polluted areas, so questions of responsibility are structurally more involved. Studies in the UK have examined this question and derived mixed conclusions. Stevenson *et al.* (1998) take the example of London and argue that a separation of responsibility and impact does exist because the poor air quality experienced in the most deprived areas of inner city London is the responsibility not of the people living in those areas – who have low levels of car ownership – but of those commuting in and out of the city to more wealthy suburbs and outlying towns. They refer to this as a 'clear injustice' built out of the three elements – exposure, vulnerability and responsibility. Mitchell and Dorling (2003) provide a more sophisticated analysis for Great Britain as a whole. They estimate total ward NO_x emissions from cars owned by residents and conclude that there is *no* significant trend related to deprivation. This is because whilst affluent wards have high rates of car ownership and use, their vehicles tend to be much cleaner than the older and more polluting vehicles characteristic of more deprived wards. However, they also identify a series of wards that were amongst the poorest in Britain where car ownership and emissions are very low, but where levels of NO_2 are amongst the highest observed. So they conclude that while the poor, in general, do contribute to the worsening of air quality (they are both producers and sufferers), the *very* poorest who live with the worst air quality have contributed the least to air pollution emissions – a situation which they conclude is 'patently unjust'.

On the other side of the world in Christchurch, New Zealand, Pearce *et al.* (2006) have examined the distribution of domestic emissions of pollution and the communities that are exposed to this pollution. They use geographically detailed estimates of pollution calculated from an atmospheric dispersion model to map both average and peak levels of particulate matter from the use of coal and wood for domestic heating, which becomes a significant burden in winter months. They conclude that both average and peak pollution loads are significantly higher among more deprived, low income communities and also that the groups responsible for producing a large proportion of the pollution in Christchurch are not the same groups that are exposed to high levels of particulates. Those areas with a large proportion of wood burning (contributing most to particulate pollution) were exposed to the lowest levels of domestic and total air pollution. They concluded that 'the discrepancy between pollution production and exposure provides evidence that environmental injustices could be operating in Christchurch' (ibid.: 935).

Justice in the air

Studies that have analysed questions of responsibility in this way move us more directly towards judgements about injustice, the third row of Table 5.1. Relatively little of the academic literature on air quality inequalities systematically tackles the justice question. Most of the work we have reviewed has been descriptive of the social distribution of air quality, and sometimes also of vulnerabilities and responsibilities. Few studies have then taken the step of making reasoned normative judgements about why these distributional inequalities matter – whether injustice exists and what should be done about it. In contrast, environmental justice activists have articulated justice arguments far more readily, typically either in the form of broad claims about the injustice of biases in pollution exposure (for example, see Faber 2008: 30) or in relation to specific local cases and circumstances, and have demanded action from air quality management authorities to redress these. As we shall see, whilst their claim-making has focused primarily on matters of distributive justice, dimensions of procedural justice and justice as recognition have also been involved.

Distributional justice and the right to clean air

Looking across examples of distributive justice arguments, these typically centre first on the threat to health and the explicit or implicit assumption that everyone has a right to breathe air that is not harmful. This is different from a notion of simple equality in which measured air quality is somehow equally shared across space and across population groups – a notion that in any case makes little sense given that the chemical make up of the air *will* always vary with patterns of air movement, dilution, dispersion and so on, and its purity is not absolute. Rather, the notion of a 'right to clean air' is effectively focused on achieving a minimum standard of air quality (or in fact a 'right to clean *enough* air'). Seeing justice in terms of minimum standards accepts and allows for variation across space, as judgement is only exerted when this variation transgresses a threshold. In other words, it is fine for some people to breathe essentially unpolluted air and for others to breathe more polluted air, but it is wrong and unjust if the 'more polluted' exceeds a threshold level of harm, particularly where it does so system-atically for some social group more than others. Common minimum standards or objectives of measured (or modelled) air quality are something that is entirely familiar to the world of air quality management. Health standards (usually based on those advised by the World Health Organisation) widely exist, specified for different pollutants, threshold concentrations and time periods of measurement. What then matters, in distributional justice terms, is where, how often and for whom these threshold measures are exceeded.

If we look back at some of the data analysis discussed earlier for England and Wales and follow this reasoning, we can find ways to discriminate between different forms of unequal air quality. The U-shaped distributional curves in Figures 5.2 and 5.3 showed that both the most deprived and the most wealthy live with above average pollution levels. But this pattern of inequality does not trans-late directly to claims of injustice, if we see the object of justice as the mainte-nance of threshold standards. What really matters is the data on exceedences of standards, which, as shown in Figure 5.4 for England and for NO_2, are strongly skewed towards deprived areas. In Wales, however, the data showed no exceedences of the NO_2 standard at all.

In Wales, therefore, and on this basis we would sustain no injustice argument – there are inequalities but no injustice, as the standard has not been breached. In England we could sustain a claim of injustice, not only because exceedences exist but also because exceedences are not being shared equally across the population of England. People who are already disadvantaged are taking the burden of unac-ceptably poor air quality on behalf of a society that is failing to control pollution levels to a sufficient degree. From this base, arguments can be, and often are, made that add further dimensions; the poorest people who are experiencing these exceedences are contributing less to the air quality problem than others (see the earlier discussion on responsibility), are most vulnerable to health consequences and are least able to avoid them. In the latter respect, Walker *et al.* (2003) identify a series of 'pollution-poverty' hotspots in England which they argue should on justice grounds be the focus of regulatory attention for reasons that particularly

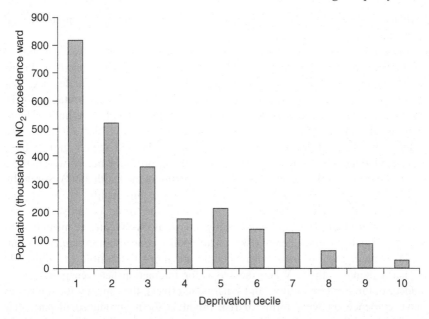

Figure 5.4 Distribution of ward mean nitrogen dioxide exceedences for England, 2001.
   Source: Walker *et al.* (2003).

Notes:
1 Annual mean standard is 40 ug/m^3, applied as a ward average.
2 2.51 million people are in an NO_2 exceedence ward, 5.1% of the population of
   England; 53% of all person exceedences are in the most deprived quintile.

centre on issues of constrained choice and capacity to relocate. Tellingly, they contrast the situation of those in poor communities with that of the Queen in Buckingham Palace:

> We are most concerned with the most deprived wards, as residents here are
> much more constrained (economically), in their choice of residential loca-
> tion, and hence unlike their more affluent counterparts, are not able to flee
> the poor air quality, or trade it off against other benefits of that location. By
> way of illustration, consider the Queen. Whilst at Buckingham Palace, she
> is resident in the ward with the third worst air quality in England (excluding
> unpopulated City of London wards). However, she trades off this cost against
> the benefits of living at the palace, and is also economically able to relocate
> to areas with much better air quality (which she does do for some of the
> year – e.g. to Balmoral in the Scottish highlands).
>
> (ibid.: 104)

Approaching distributional justice in terms of a right to a common minimum standard of air quality therefore appears to make much sense and appeals in abstract logical terms. But the reality is more problematic. Two significant deficiencies in relying on standards have been identified by environmental justice activists.

First, whilst air quality standards are supposed to be set to take account of particular sensitivities and to protect the most vulnerable, their capacity to do this is at least uncertain, if not deeply constrained. There are considerable uncertainties in the science by which standards are set and risks assessed (Montague 2004), leading to major disagreements as to whether or not a simple threshold of harm can be determined. Epidemiological evidence has suggested that adverse health effects do occur at levels below current standards, and expert groups have conceded that there is no precise concentration threshold below which there are no adverse health effects of exposure (Department of Health 1998). Similarly, whilst the significance of the interaction between air quality, ill health and the social and place dimensions of people's lives (the 'triple jeopardy' discussed earlier) is still being debated scientifically, this could well logically mean that, to protect 'equally' against health risks, an air quality concentration standard would need to be tighter in areas where people are more susceptible or vulnerable to harm than in others. When the scope for interaction between pollutants is added, in the form of additive and synergistic effects, the capacity for science to have evidence on what forms of risk might be being produced in particular localities and for particular bodies is even more constrained. Brown *et al.* (2003) comment on how, for community work on asthma incidence by 'Alternatives for Community and Environment', an environmental justice group based in Roxbury, Boston, this means that campaigning has to focus not just on scientific evidence but also on wider 'narratives of ill health' and on political perspectives that 'situate asthma in terms of housing, transportation, neighbourhood development, the general economy and government regulations' (ibid.: 460). Relying on the narrow and reductionist science of air quality standards is, by these arguments, insufficient.

Second, using standards and data on where these are being breached, to make justice judgements assumes good knowledge of the real world variation in air quality and air pollution exposure. However, the capacity of air quality and regulatory agencies to fully know air quality, to measure its complex variation in different circumstances, is highly constrained. As we have seen, there is a patchwork of knowledge on air quality at different scales of abstraction and aggregation, knowledge that might appear relatively substantial and complete in some places but elsewhere is deficient, if not absent. For this reason, just because available monitoring data or air quality models do not reveal exceedences of standards, this does not necessarily mean they haven't happened – in particular places, or over time periods not accounted for by measurement or modelling protocols – and had consequent impacts on people's health. Environmental justice activists have also argued that there may be systematic biases at work such that the scrutiny, monitoring and prosecution of major industrial polluters is not necessarily carried out in an even-handed way. Laxer approaches, it is claimed, are adopted for polluters in low income and minority areas, and tighter ones in politically influential middle-class communities. Such claims have been scrutinised statistically, producing both supportive and less supportive evidence (Gouldson 2006; Heiman 1996; Mennis 2005; Ringquist 1998), but the suspicion that pollution monitoring and regulation are not even-handed has been enough to raise questions about the wisdom of relying on government authorities and scientific experts to assess and address air quality problems.

Procedural justice, science and grassroots action

Such concerns bring us to the arguments deployed by community activists that centre on procedural justice as well as on matters of distribution. These justice claims make the case for local participation in the decision-making processes that scrutinise the scientific knowledge base, set standards and decide on the forms of intervention measures to be taken to deal with air quality problems – in essence, for those at risk to be inside rather than outside of the processes of assessment and knowledge production that feed into air quality decision-making, regulation and enforcement. Whilst these arguments are made on the principled grounds that there is a specific community of justice involved that has a right to have its voice heard and to be recognised (see Chapter 3), they are also born out of the critiques of science and its application that have been outlined above. Activists point to the need for the scrutiny of institutional knowledge and understanding, rather than taking what is available, and officially recognised, as sufficient and reliable. As Azibuike Akaba (2004: 22) from 'Communities for a Better Environment' states, 'science is often viewed as information given to us by community agencies: we are just supposed to accept it, as opposed to questioning it or generating it ourselves'. Following this line of reasoning, he argues criteria for injustice cannot rest solely with a device, such as an air quality standard, that is dependent on uncertain science for its existence and on incomplete monitoring and enforcement for its power. Inequality is seen as much within science and regulation – it has 'rewarded polluters' in Akaba's terms – as it is within historically produced social and environmental conditions.

For these reasons, environmental justice activists have increasingly taken up initiatives to draw in and develop sources of expertise that support their problem framings and have advocated alternative decision rationales to those of main-stream air quality management and health protection. Asthma has figured at the centre of many such instances of local health politics, particularly in the US (Brown *et al.* 2003). Sze (2007) analyses the 'asthma activism' which was impor-tant to environmental justice organising activity in New York through the 1980s and 1990s, for example, campaigning by the South Bronx 'Clean Air Coalition' against a new medical waste incinerator, and West Harlem Environmental Action's (WE ACT) protests against a new diesel bus depot, which were both in areas that had the highest hospitalisation rates for asthma in the city. Corporate and municipal supporters of these, and other, proposed polluting facilities argued that there was no proof that childhood asthma would be worsened by pollution emissions. Activists countered that official risk assessments failed to take account of cumulative exposure to pollutants and the fragile health status of many young and old residents, and that a precautionary approach should be taken. WE ACT also developed a programme of collaboration with nearby Columbia University's School of Public Health to undertake air pollution measurement at key traffic intersections, and to pursue research into how childhood asthma is linked to patterns of pollution exposure in low-income communities both before and after birth (Brown *et al.* 2003). See Figure 5.5 for one of the maps of childhood asthma

Figure 5.5 Childhood asthma hospitalisation patterns and transport and waste facilities in Manhattan.
Source: West Harlem Environmental Action and the Columbia Centre on Children's Health.

hospitalisation rates produced as part of WE ACT's work with the Columbia University centre.

Another example of participatory processes being used to investigate air pollution problems has focused around so-called 'bucket brigades', where use is made of a low-tech, low-cost technology, first developed and used in California, to collect air samples for laboratory testing. As well as being deployed in many places around the US, environmental justice groups in South Africa have made use of the bucket brigade community-based monitoring system (Kalan and Peek 2005), and training and monitoring programmes have been undertaken in Kenya, Mexico and Australia (see www.gcmonitor.org). One of the arguments for such community-based monitoring is that official monitoring networks do not take account of the variability of local conditions over space and time. This has also pushed some environmental justice activists towards considering community involvement in the use of various biomonitoring and genomics technologies which more directly assess the overall burden of pollutants on specific individual bodies. There are though very mixed views about the ethical and political implications raised by the use of such technologies (Shostak 2004; Sze and Prakash 2004).

As well as focusing procedural justice claims on scientific and governmental processes, some community groups have also sought to set up processes of involvement with industrial operators through 'Good Neighbour Agreements' (Illsley 2002; Lewis and Henkels 1998). These agreements provide a formal and sometimes legally backed mechanism for residents to negotiate the standards of operator performance that are to be maintained in terms of, for example, emission levels, access to information and the handling of complaints. As such they constitute an attempt to instigate a particular, locally determined set of distributional and procedural justice principles that recognise the local community as a legitimate partner in decision-making processes.

Summary

We have seen that air quality is both a compelling case for exploring questions of inequality and a clear example of the multidimensionality and complexity that can be involved in making claims about environmental justice. It has been emphasised throughout though that claims about air quality are contested and that the status of science as an arbiter of reliable evidence can be fundamentally at issue. We have seen that even undertaking a first step of generating evidence on the social distribution of air quality is problematic. Choices have to be made about the forms of pollution and social difference to be concerned with, the scale at which analysis is to be undertaken, and so on. In reviewing some of the studies from around the world, but in particular those from the US and the UK, we found a diversity in outcomes that reflected in part these choices, but also the ways in which patterns of air quality and social difference are contextualised within the particularities of places and their histories. However, it is not just a matter of whether there is more or less air pollution in areas that are more or less wealthy, poor or racially distinct. There can be different vulnerabilities to health impacts amongst population groups and forms of double or triple jeopardy which mean that consequences are all the more severe and accumulative for some people than others. Establishing the nature of such interactions between air quality, social context and health is leading to ever more sophisticated methodological designs, but also to community-based processes.

Responsibility for the production of air pollution can be clearly assigned to industrial and corporate actors, but it might also be more widely distributed amongst the population, particularly in relation to car driving. Injustice claims can be concerned about discontinuities between who is responsible for, and benefiting from, activities generating air pollution and who is suffering the consequences, but can also be made on other distributional and procedural grounds. One approach is to conceive justice in terms of the enforcement of equal minimum air quality standards, but problems with how well protective standards can be specified, particularly for vulnerable populations, the uncertainties of science and the limits of monitoring and regulation can all undermine how a standards rationale is viewed by environmental justice advocates. Participatory processes of community-based monitoring and investigation have therefore been pursued as an alternative bottom-up approach to generating evidence of the everyday experience of poor air quality and impacts on health.

Further reading

An accessible introduction to urban air pollution problems is provided by Elsom (1996) and in more technical detail by Colls and Tiwary (2008). Methodological issues in undertaking socio-spatial analysis of air quality are discussed in some detail in Liu (2001) and Jerrett *et al.* (2001). The relationship between air pollution inequalities and health outcomes are explored in special issues of *Health and Place* (2007, vol. 13, no. 1) and the *Journal of Epidemiology and Community Health* (2004, vol. 58). Papers by Briggs *et al.* (2008), Pastor *et al.* (2006) and Grineski (2007) provide examples of the sophisticated research designs that are now being applied to link patterns of air quality to health outcomes.

For an excellent analytical account of environmental justice activism on air quality and health issues see *Noxious New York* by Julie Sze (2007) and Brown *et al.* (2003) on both New York and Boston. The evolving use of the 'bucket brigade' can be tracked at www.gcmonitor.org. A special issue of the *American Journal of Public Health* (2009, vol. 99, part S3) covers a wide range of health issues and processes of community research: http://ajph.aphapublications.org/content/vol99/issueS3/. To follow up on the complex ethical debates raised for environmental justice by biomonitoring and genomics technologies see the discussion in Shostak (2004).

6 Flood vulnerability

Uneven risk and the injustice of disaster

Flooding presents a threat to millions of people around the world. It kills, injures and causes long-term health impacts, creates severe damage to property and livelihoods and disrupts the functioning of whole urban and rural systems. It does so very unevenly though and, given this, flooding might appear an obvious candidate for fitting within an environmental justice frame. It is therefore striking that, for the first 20 years of environmental justice campaigning and research in the US, flooding was not within the frame of political agenda setting at all. Indeed it took the devastating impacts of Hurricane Katrina for the connections to be made between the dominant US race and poverty framing of environmental justice and the ways in which many of the poor and black residents of New Orleans experienced the flood and its aftermath.

To explain why flooding did not become part of environmental justice politics in the US before Katrina, we can look to the arguments made in Chapter 2 about the contextualised construction of collective action frames and how justice concerns and claims are grounded within a particular cultural and political time and place. Because of where it came from and how it evolved, environmental justice in the US was understood for many years as being essentially about race and poverty and pollution and waste – about human and technological rather than 'natural' forms of risk and hazard (Fothergill et al. 1999). In contrast, if we look to the UK flooding was part of a defined environmental inequalities policy and research agenda from when this first materialised (Walker and Burningham 2011). To a large extent this was because the remit of the Environment Agency – who commissioned early research projects on environmental justice and inequalities and promoted policy attention across government (Chalmers and Colvin 2005) – included flooding as a key responsibility. Flooding was in the institutional frame rather than outside of it, contextualised by a very different process of introduction and diffusion from that in the US.

However, issues of social difference, inequality and injustice were not 'discovered' only after flooding had became part of an environmental justice discourse. Far from it. There is a major stream of work on flooding and other forms of 'natural' hazard that focuses on questions of vulnerability and that seeks to understand the processes and structural factors which mean that poor and marginalised communities, women and ethnic minorities can be more severely affected in myriad ways than others (Alwang *et al.* 2001; Cutter *et al.* 2003; Pelling 1999; Pelling 2003a; Wisner *et al.* 2004). Much of this work has been focused on the developing world, initially at least rooted in political ecology frameworks (see discussion in Chapter 3), providing a powerful analysis that has increasingly become part of mainstream thinking on hazards and disasters. In work on flooding in Europe and North America the vulnerability perspective has also become important (Green 2004) and influenced disaster management practices and strategies (McEntire 2006; Twigger-Ross and Scrase 2006; Wisner 2001), including in relation to climate change (Dolan and Walker 2004).

There is therefore much to draw on in a discussion of floods, inequality and justice, and much to wrestle with. In this chapter I concentrate primarily on where an environmental justice (or closely related) frame has been used to examine flood risk and flood experience, which focuses attention on floods in the developed world, primarily the US and to some extent the UK. This, from a global perspective, narrows the scope of the discussion considerably, so at various points in the chapter comparisons will be made with the playing out of inequalities in other parts of the world.

The discussion begins with some preliminary consideration of how flooding is distinctive as an environmental justice issue. We then consider evidence of two dimensions of inequality in broader terms: evidence of who is exposed to flood risk and evidence of who suffers most severely from flood impacts. Interlinking these two dimensions is crucial, as it is a matter not just of who in society lives in risky places, but also of how vulnerable they are to flood impacts. These interrelated dimensions of inequality were introduced in Chapter 3 and explored further in Chapter 5, but the case of flooding is particularly appropriate for examining them in more depth. We then examine Hurricane Katrina and the resultant flood impacts on New Orleans as *the* case which fits most explicitly within an environmental justice frame. Building on Katrina, the chapter finishes with a discussion of the normative terms of 'flood justice' and various alternative ways in which this might be understood. Box 6.1 summarises the cross-cutting analytical themes highlighted in this chapter, moving from vulnerability through to justice concepts.

Characterising flooding: values, time and nature

As with each of the topics of Chapters 4–8, there are distinctive characteristics of flooding which need to be incorporated into our analysis. These encompass matters of meaning and value and the particularities of temporal and spatial patterns.

In Chapter 3 it was discussed how environmental features can be valued in multiple ways and be given alternative meanings. Floods are a case in point. Whilst focusing on flood as a form of disaster is to evaluate 'an excess of water in a place that is normally dry' (Arnell 2002: 112) as something resolutely bad, flooding of some types, in some places and for some people can also be considered a good thing. This is widely recognised in terms of the role of floods in

---

Box 6.1 Flooding and cross-cutting themes

By discussing flooding through an environmental justice frame we will be able to explore:

- the importance of understanding the distribution of vulnerability to harm amongst different demographic and social groups, and how this varies over and beyond the spatial patterning of who lives in hazardous places;
- the role of history and geographical context in explaining the processes that have produced contemporary patterns of inequality in flood risk and vulnerability;
- the ways in which people across different intersecting social and demographic categories can be subject to multiple and interacting forms of inequality, in the case of flooding also structured over time;
- the three core forms of (in)justice – distribution, procedure and recognition – and how these can all be found in an analysis of flood experience;
- the value in also positioning complex experiences of injustice, such as from flooding, within the 'capabilities' framework.

---

fertilising and irrigating farmland and sustaining wetland landscapes, but Mustafa (2005) also notes that in his research in Pakistan some better-off urban residents living in well built houses welcomed the recurrent annual flooding as a cleansing of the urban space of garbage and dirt. Poorer residents who lost their homes and belongings evaluated the flood in radically different terms. It is also the case that some obviously flood-prone places are still positively evaluated – coastal, riverside and lakeside spaces often have high aesthetic (and land) value (Collins 2010). There is therefore some ambiguity and potential contestation in the meaning and evaluation of floods – just as in Chapter 7 we will see that greenspace has an overtly 'slippery' normative status.

Floods are distinctive in temporal terms. They are events that happen when water moves and accumulates unusually. There is therefore a periodicity to them that makes them quite different from other environmental justice topics. The interaction between floods and people has distinctive phases, from pre-flood preparedness, to warning, evacuation (in some cases), the flood period, recovery and living with the aftermath, through to rebuilding and restitution. Matters of inequality and injustice might be at issue within and through each of these phases, relating to who is prepared, who is warned, who evacuates, who is killed, who suffers the worst health effects and damages, who is best able to recover and who gets to rebuild. The intertwining of flood and social difference is therefore dynamic and shifting over time.

Floods are also distinctive in spatial terms; they happen in particular places where water gets to flow, accumulate and occupy spaces that are 'normally dry' (Walker *et al.* forthcoming). There is therefore an important role for nature in the processes that lead to these movements of water (heavy rainfall, severe storms, snowmelt, glacial lake outbursts and so on), but this does not mean that they are solely 'natural' hazards. The determination of where floods become problematic – where they interact with and damage things that are valued – is down to human processes of establishing settlements, building infrastructure, growing crops and so on, and the ways that water flows are channelled, directed and resisted. The transition through a flood becoming a hazard and then becoming a *disaster* is even more resolutely about the social processes that create vulnerability and achieve different degrees of coping and resilience. Therefore questions of decision-making, responsibility, power and the role of the state in protecting citizens from harm are just as relevant as for other environmental justice topics – floods are not just 'acts of god'. So environmental justice research, analysis and claim-making have as much relevance to so-called 'natural hazards' as to more obviously man-made forms of environmental harm. This critique of the notion of natural hazards or disasters has informed much existing work concerned with understanding patterns of inequality in hazard, and specifically flood experience, and it is to this literature that we next turn.

Inequalities in flood exposure: who lives with flood risk and why?

A basic starting question for an exploration of inequality and flooding is to ask: who is living at risk of flooding? As the risk of flooding is focused primarily on particular spaces in proximity to rivers, coastlines and other water bodies, who is living within such 'at risk' spaces – a broad and even profile of the population or particular types of people with particular social characteristics?

If we looked to work undertaken in developing countries, particularly within political ecology frameworks (see Chapter 3), much experience and empirical evidence show that those who have limited livelihood options end up occupying the most hazardous environments. They live where they can and where they are pushed to by market forces such that 'it is no accident that a major slum in San Juan (Puerto Rico) is frequently inundated by high tide, that the poorest urban squatters in much of Asia live in hazardous floodplains' (Susman *et al.* 1983: 277). This is widely accepted as deeply and systemically problematic for developing country contexts, although whether marginalisation processes always and everywhere operate in this way has been challenged (Kates and Haarmann 1992). We might speculate though as to whether conditions and processes producing flood hazard exposure could be different in more developed country contexts, such that alternative social geographies of flood risk might also exist.

Evidence is partial and incomplete but a few studies from the US and UK can give us some insight into the variation and complexities involved. These are all examples of a 'pre-flood' approach to generating evidence on flood potential – see the discussion in Chapter 3 – rather than a 'post-flood' approach, which we will use in the discussion of the Katrina flood.

In an analysis of the relationship between areas of estimated flood hazard and race in New York, Maantay and Maroko (2009) conclude that although there are instances of significantly disparate risk in particular areas of the city, there is no consistency between boroughs regarding the risk of flood exposure and race. Looking at processes operating over time, they conclude that this is because of a number of differentiating factors:

> variations amongst the boroughs' historic patterns of residential settlement; different levels and chronology of industrial development along the water-front; recent and historic landfilling of coastal wetland areas, which subse-quently enabled development at different times; recent de-industrialization efforts and gentrification in certain areas of the city; and cultural changes over the years concerning the desirability of living along the waterfront and therefore the flood zones.
>
> (ibid.: 11)

Moving out to a bigger scale of analysis, Ueland and Warf (2006) cover 146 cities in the US South, but rely on a crude proxy for flood hazard potential – the altitude of the land on which people live. Their particular interest is in looking for racial disparities in population patterns to test 'whether historically disenfran-chised African-American communities have been relegated to low-lying topo-graphic areas in their respective cities while their white counterparts have commandeered higher elevations' (ibid.: 53). Their results reveal strong patterns in which whites do typically occupy higher-altitude locations. The only excep-tions were some low-lying coastal or riverine cities in which property values were inversely related to altitude and white populations lived at lower elevations in high value properties along the coast and river fronts. These can be seen on Figure 6.1, which maps the outcomes of their correlations for the 146 cities. Through a series of case studies they illustrate the urban dynamics that have produced these contemporary patterns, tracing, for example, the racialised state and market processes in Washington DC that concentrated black housing along the Anacostia River; the combined impacts of racism in labour and housing markets in Mobile that pushed poor and black populations onto poor quality, low lying coastal land near to heavy industry; and the reverse outcome of underlying segregation in housing markets in Ft Myers-Cape Coral, Florida, leading to whites occupying high value waterfront properties. So whilst finding variation across these urban contexts, they conclude that given the racialised history of the southern states:

> The discriminatory dynamics of housing and labor markets, including the lower average incomes of blacks and prejudicial access to mortgage lending

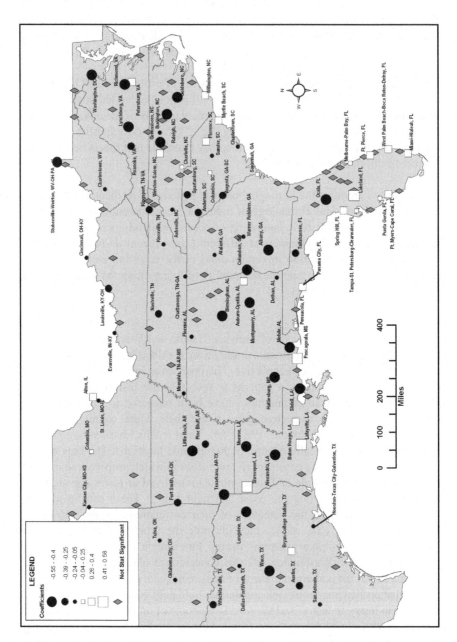

Figure 6.1 Correlation coefficients between percentage of black and mean altitude in urban areas of the US South.

Source: Ueland and Warf (2006).

and exclusionary zoning, have operated formally and informally for decades to generate geographies in which African Americans find themselves disproportionately concentrated in low-lying areas.

(ibid.: 73)

In the UK several national scale analyses of who is at risk from coastal and river flooding have been completed (Fielding and Burningham 2005; Walker and Burningham 2011; Walker *et al.* 2003; Walker *et al.* 2007; Werrity *et al.* 2007), using official Environment Agency maps of flood risk zones. The concern in these studies has been largely with patterns of social class and deprivation rather than with race. Walker *et al.* (2007), in the most substantial and disaggregated study for England, found a strong socially regressive gradient, such that people who are strongly deprived are more likely to live in a flood risk area than those who are much less deprived. On disaggregating between river and sea flooding zones though, they found that the overall profile observed in the aggregated data was entirely created by the pattern within the sea flooding zones (itself a lesson in how descriptive claims of association are dependent upon how categories are delineated and aggregations and disaggregations performed). The profile for river flooding (Figure 6.2) is very flat, with little variation from most to least deprived. However, for sea flooding (Figure 6.3) the profile is very different, with a strong concentration towards the most deprived deciles. The total population at risk in deciles 1 and 2 is approximately 750,000, compared to only 80,000 in deciles 9 and 10, the least deprived.

Explaining why these different patterns exist involves looking again to some specifics of geography and history and to a set of interacting social and

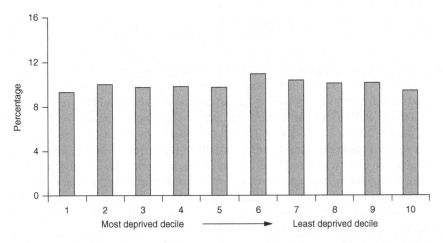

Figure 6.2 Percentage of total population within flood risk zones for river flooding by deprivation decile.
Source: Walker *et al.* (2007).

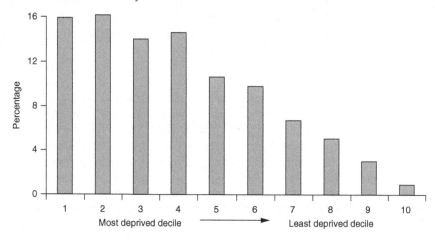

Figure 6.3 Percentage of total population within flood risk zones for sea flooding by
deprivation decile.
Source: Walker *et al.* (2007).

economic processes. The fairly even deprivation distribution along river flood
plains suggests that low lying flood risk land is not being avoided by those with
higher incomes but has amenity and property value, linked in part to the many
historic settlements located on rivers that regularly flood (Oxford, Tewkesbury
and York, for example). On the coast, however, three factors are likely to have
contributed to very different population patterns. First, the UK has a series of
old industrial ports which developed with a largely working-class population
living nearby, including, for example, the ports of London, Hull and Liverpool.
Some of these port cities have a large part of their urban area within the coastal
flood plain. Second, within such cities and other coastal towns the strategies
of housing developers, reflected also in land prices, have meant that poorer
quality housing tended to be built on lower lying land, sometimes originally
marshland which was reclaimed for development. Better quality housing, as in
Victorian and Edwardian Britain, was often to be found on more elevated land
(Daunton 2000) – in line with Ueland and Warf's hypothesis for the US, but
along class rather than racial lines. Third, there is a series of seaside resort towns
around the coast which had their heyday in the first half of the twentieth century,
providing holiday destinations for the mass of the British public. Many of these
resorts have since declined (Agarwal and Brunt 2006), faced with competition
from destinations abroad, and have become increasingly deprived, suffering from
serious associated social problems.

The variety of processes that can operate to push or pull populations of different social status towards hazardous places is examined particularly carefully by
Collins (2010). He builds on the work of Davis (1998) to show how hazardous

places can simultaneously be imbued with both negative and positive values, so that elite groups can be found in amenity rich, hazard-prone locations such as in wildfire areas and on beach and river fronts. His own qualitative empirical study examines exposure to flood risk in socially elite and socially marginal communities affected by a flood in 2006 in Paso del Norte, which spans either side of the US–Mexico border. He explains how flood prone areas along the river include many of the region's most economically and politically powerful residents and also some of the most marginal and socially vulnerable, with highly differentiated urbanisation processes at work. On the one hand, low paid labourers seeking to establish their livelihoods in the region's rapidly expanding industry have been channelled into flood prone informal settlements; while, on the other, richer residents have moved into high quality suburban developments along the river on the US side of the border, supported by state infrastructure investments. When the flood happened in 2006, the impacts were felt far more severely in the poor areas, as those with more resources, status and recognition, and with more influence over decision-making, proved much better able to cope and recover.

This clearly indicates that focusing on inequalities just in terms of 'who lives in the flood plain' can only take us so far. It is undoubtedly a crucial entry point to experiencing flood impacts, but in the event of a flood much else then comes into play which differentiates between social groups *within* the flood-plain population – which, as we have seen, may not *always* only consist of the most economically and politically marginalised. This brings us to the examination of inequalities in vulnerability.

Inequalities in vulnerability: who suffers flood impacts?

That there is more to vulnerability to flood impacts than just location is clearly conveyed in Figure 6.4. This provides an integrated view of disasters that brings together much more than 'location relative to hazard', which can be found at the far left of the diagram. In particular, this schematic includes 'resistance', the capacity of an individual or a group to withstand the impact of a hazard, and 'resilience', the ability to cope with or adapt to hazard stress, encompassing preparedness and processes of relief, rescue and recovery (Pelling 2001, 2003b). All of these elements, a political ecology perspective emphasises, are conditioned by the degree of access to rights, resources and material assets that people can secure, which is closely linked to the dominant social divisions (race, class, gender, age) in a particular society.

Again, there is much documented evidence from developing world contexts which shows that whilst some positive stories can be told of households having strong social networks and capacities to adapt and endure, there is no doubt that floods have the capacity to sweep away the livelihoods, homes and lives of those on the margins and without access to financial and related resources in an extreme and devastating way (Bankoff *et al.* 2004; Few 2003). The recent (2010) massive flood in Pakistan is a case in point; it is estimated to have killed over 2,000 people and displaced more than 21 million across a vast swathe of the country.

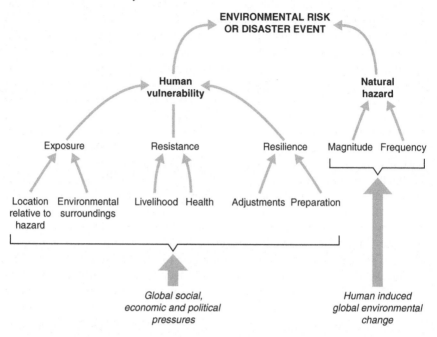

Figure 6.4 An integrated view of disasters.
    Source: Adapted from Pelling (2001).

In developed world contexts poverty also plays directly into patterns of vulnerability, resistance and resilience during and after disasters, albeit typically on a lesser scale (Fothergill and Peek 2004). For example, access to flood insurance is strongly related to the availability of financial resources unless some form of universal provision or state support is available. Research by the Association of British Insurers (2002) found that 50 per cent of households in the lowest income decile in the UK do not have contents insurance. Whyley *et al.* (1998) also show that uninsured households are disproportionately likely to have few savings and to be facing financial problems. Deprived areas with high crime rates will also have high insurance premiums, making insurance even more likely to be unaffordable for residents (Ketteridge and Fordham 1998). Research on a major flood in Hull in the UK (see Box 6.2) showed strong differentiation between the experiences of households with and without insurance. This and much other research has also shown that those on low incomes find it hard to cover the costs associated with evacuation and temporary accommodation. Financial resources act as a 'buffer' to flood impacts, and those without such a buffer will be hit harder by the various impacts of flooding. In reviewing evidence of the relationship between flood inequalities and deprivation, Walker *et al.* (2007: 45) conclude that 'while it is important to emphasise that not all vulnerable individuals and households are

Box 6.2 The 'forgotten city' of Hull and the 2007 flood

In July 2007 a series of floods afflicted towns and cities across the UK, including Hull, a port city in the north-east of England. It became known as the 'forgotten city' because of the way media reporting overlooked this peripheral city, the ninth most deprived district in England. The flood happened because of the failure of the drainage system to cope with extremely heavy rainfall. Over 8,500 houses were flooded, affecting over 20,000 people or 8 per cent of the population. No specific flood warning was given, as the warning system linked only to predicted sea or river levels which were not expected to be problematic. Of the flooded households, 6,300 were forced out of their homes and over 1,400 people lived in caravans for an extended period.

Figure 6.5 Post-flood caravan living in Hull.
Source: Rebecca Whittle.

Given that the city as a whole was so deprived compared to the rest of England, it proved particularly vulnerable to the impacts of the flood. A database collated by the city council showed that 45 per cent of those flooded fell into the most vulnerable category comprising residents over 60 years of age, people with disabilities and single parents with at least one child under five. In two longitudinal qualitative studies carried out after the flood, involving diary writing and interviews with adults and children in flooded households, the severity, differentiation and extended nature of the flood impacts became clear (Walker *et al.* 2010; Whittle *et al.* 2010). For example, for those people without household contents insurance the experience was of losing everything even temporarily touched by flood water because of contamination with sewage:

> When the council told us to throw everything out which had come into contact with the water as the water was contaminated we were absolutely gutted – more so when the dust cart came and took it all away. As we weren't insured all we could think of was "Oh my god we've got nothing left and it took us years to get that sort of furniture together".'
>
> (Marion, diary)

Those needing accommodation to live in could find it enormously difficult, in part because of local estate agents pushing up rents and making profits 'from others' misfortune'. Even for those people with insurance it took extended, frustrating battles with loss adjusters to settle a claim, a task much harder for those without the time, confidence or capability to take this on. Health impacts could be severe, with high levels of stress produced by trauma, anxiety, disruption and the pressure on family relationships (Sims *et al.* 2009). The diaries and interviews revealed many such instances:

> Some days I just felt like jumping off the Humber Bridge. It's been that low, it's been that bad, except I'm not brave enough to do it. But the state of mind you've been in – some days I've just sat in here and just sobbed and sobbed and sobbed.
>
> (Leanne, interview)

Older people were a specific concern for the council, but it proved difficult to identify all cases of need and some people were reluctant to be helped or even to admit to having been flooded – they were worried about having to move out of their secure home environment and never being able to return. People with disabilities could find the stripping out of the interior of their houses, especially when adapted for their disability, and moving into alternative accommodation particularly difficult.

deprived, it is nonetheless true that deprived neighbourhoods contain concentrations of vulnerable individuals'. Other factors that add to vulnerability apart from being deprived are identified as including old age, having pre-existing health problems or a disability, being female, being from an ethnic minority and being a single parent with young children. Where these intersect across 'multiple levels of social life' (Buckle *et al.* 2000), such as within deprived communities, there can be particularly intense problems in coping and recovering.

Part of the capacity to cope and recover relates to processes of discrimination – or lack of recognition – which have been documented as featuring in many disaster experiences. Those who are poor are often also those who are discriminated against – and vice versa – given the way that distributional inequalities and lack of recognition interact (see Chapter 3). A wide-ranging review is provided by the 2007 *World Disasters Report* (International Federation of Red Cross and Red Crescent Societies 2007). This takes a broad view of discrimination to include race and ethnicity, gender, age, migrant status, disability, caste and indigenous status, amongst others. Within an overall framework asserting that 'disasters don't discriminate, people and governments do', and that fundamental human rights are as relevant to disaster situations as any others, varied instances of discrimination specifically in relation to flood experiences are described. For example, after major floods in Romania in 2005 media reporting stigmatised the Roma ethnic group as spreading infectious diseases, and rehabilitation programmes supporting the rebuilding of houses excluded them as they lacked formal papers establishing rights to property and land. In Kenya repeated failures by the government to protect pastoralist communities from the impacts of drought and flooding reflect an institutionalised rejection of pastoralism as a viable and legitimate way of life. In Bangladesh, survey work found that people with disabilities (6 per cent of the population) were widely excluded from flood relief and rehabilitation programmes because of accessibility problems and the difficulties in providing for their particular care needs. Family networks through which caring was shared could be severely disrupted by floods, making people with disabilities especially vulnerable.

All are not, therefore, equally vulnerable to the impacts of flooding. Even from this rather rapid and selective review we are left with a complex picture of the patterns and processes that interact before, during and after flood events to make them disasters for some and little more than inconveniences for others. This picture can be further filled out by examining one case in detail, not necessarily the most typical but certainly the most researched and the most talked about within an environmental justice frame.

New Orleans and the Katrina flood

As noted at the beginning of this chapter, before Katrina in 2005 flooding had rarely been a focus for environmental justice activism in the US. That is not to say that the class and racial differentiation of disaster experience in the US was entirely unrecognised. For example, Fothergill *et al.* (1999: 169), in looking across a large number of case studies of hurricanes, floods and earthquakes, found 'a picture of increased vulnerability and risk to disasters for racial and ethnic communities'. However, there had been nothing as immediately shocking or as dramatically divisive in its playing out as the case of Hurricane Katrina, and it touched a political nerve that immediately resonated with the racial politics of the environmental justice community. The Katrina story, therefore, bears telling in some detail, moving across and beyond the dimensions of flood inequality discussed up to this point.

In the early morning of 29 August 2005 Hurricane Katrina made landfall on the Gulf Coast. The surge of water pushed onto the coast by storm winds did immediate damage along the coast but, funneled by the wind, also overwhelmed the canals and artificial levees intended to protect the low lying and heavily populated urban space of New Orleans. The levees should have been able to cope with the impacts of a storm of the intensity of Katrina, but they were poorly designed and inadequately maintained (Freudenburg *et al.* 2009). As the levees failed, water flooded into the lives of hundreds of thousands of New Orleans residents. A few neighbourhoods were little affected, others submerged by up to 15 feet of water. What then transpired for the city's residents has been researched from many different perspectives. Evidence has been collected in many different ways – from census analysis to large scale surveys and personal narrative accounts – and is not unproblematic or uncontested (Sharkey 2007), some seeing significance in disparities that others dismiss; for example, see Block (2006) for a libertarian perspective. Environmental justice framings have been taken up in this work, alongside many others that seek to position the Katrina experience in relation to particular ways of understanding and analysing what transpired. A thorough review of this still growing body of research is beyond the scope of this

Figure 6.6 Inundated areas in New Orleans following breaking of the levees during Hurricane Katrina. New Orleans, Louisiana, 11 September 2005.
Source: Lieut. Commander Mark Moran, NOAA Corps, NMAO/AOC.

chapter, but some of its key dimensions can be outlined, beginning with the socio-demographic profiles of the parts of the city and wider region that were submerged by flood water and those that were not. We then examine uneven patterns of vulnerability across the different phases of the disaster experience.

Patterns of flooding

Several analyses of the impact of Katrina provide data on the socio-demographic profiles of the affected areas (based on official damage mapping and census statistics), disaggregating the population in various ways. Along the wider stretch of the Gulf coastline 644,000 people lived in heavily damaged areas and 90 per cent of these were flooded (Logan 2006). The flooded population was poor by US standards, 20.7 per cent living below the poverty line compared to 12.4 per cent across the whole US population (see Figure 6.7). About a quarter of this affected population were children, with a child poverty rate almost twice that for the nation. Racial differences were even greater: 46 per cent of those affected along the coastal region were African-American, compared to 12.9 per cent nationally.

Focusing in on New Orleans, about three-quarters of the population of the city – 354,000 people – lived in heavily flooded areas (Logan 2006). Of these, 75 per cent were African-American and 29 per cent officially living in poverty, compared to 46 per cent African-American and 25 per cent living in poverty in other unaffected parts of the city. As shown in Figure 6.8, the five worst affected neighbourhoods of the City included Lakeview, which is almost entirely white and prosperous. But this sits alongside Mid City, East, Gentilly and Lower 9th wards which had black population proportions ranging from 70 to 96 per cent and corresponding poverty levels as high as 44 per cent.

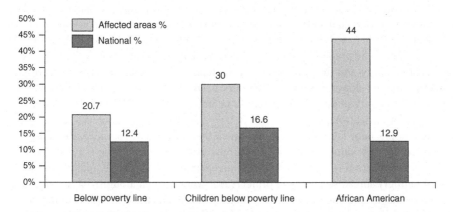

Figure 6.7 Poverty and child poverty rates and percentage of African-Americans for areas across the region affected by Hurricane Katrina compared to national data.
Source: Adapted from Gabe *et al.* (2005).

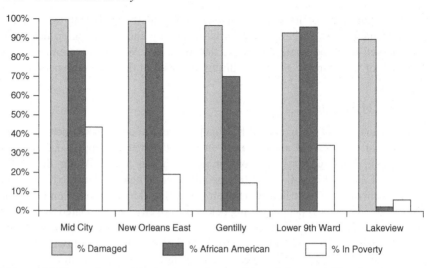

Figure 6.8 Comparisons of the five worst damaged New Orleans neighbourhoods.
         Source: Adapted from Logan (2006).

In broad terms some of these disparities are not surprising. The three coastal states affected were amongst the poorest in the country, and the level of poverty amongst the most affected population reflected pretty closely the average level of poverty in the region. In other words, the affected population was not especially poor in regional terms, even though it was when compared to the national average. However, the bias towards African-American populated areas, in New Orleans in particular, is stronger than would be simply explained by the high background population and therefore needs a more considered explanation.

Morse (2008), in a report taking an environmental justice perspective on the disaster, traces the history of the development of New Orleans in order to understand how the lowest lying areas and, in the event, less well protected areas of the city came to be predominantly African-American. Looking back to the late nineteenth century and the end of slavery, he traces how a segregated racial geography emerged along 'classic southern' lines. The dominant white population living on high quality plantation land on the natural levees of the Mississippi selected areas for African-Americans to live in that had various disadvantages – air pollution, noise, poor facilities, flooding – such that they were geographically marginalised on 'low value, flood prone swamplands at the edge of the City' (ibid.: 6). Such segregation was reinforced by slum clearance in the 1930s and the building of publicly subsidised housing projects that were divided along strict racial lines. Four 'black' housing projects were located in low-elevation positions in the back of town, and two 'white' projects on higher land towards the front of town. Over time new areas of housing were established that were occupied predominantly by better-off white populations. Some of this new housing was on reclaimed swampland near to Lake Pontchartrain, which in 2005 was badly flooded (the Lakeview district in Figure 6.8). But predominantly, new transport routes opened up a widespread process of 'white flight' into suburban sub-divisions, leaving behind an increasingly poor and black urban core population. This unevenness in mobility was again significant when the flood waters threatened in August 2005.

Evacuation and death

The day before Hurricane Katrina arrived on the Gulf coast, a mandatory evacuation order was announced by Mayor Roy Nagin. This prompted a mass evacuation from the city, although many were left behind. Most of the evacuation took place by private car, but over one in three African-Americans had no vehicle, rising to 60 per cent of those that were poor (Morse 2008). Accordingly, some of the worst affected parts of the city were those where people had no immediate means to evacuate and were poorly connected to the major highways out of the city. Attempts by the city authorities to organise evacuation by bus were grossly insufficient and poorly organised. Both this failure and the overarching emphasis on private cars as a means of mobility, critics argued, reflected recurrent discrimination against minorities and the poor in US transport provision (Bullard *et al.* 2009). Consequently, when the flood waters broke through the levees, an estimated 70,000 people were still in the city, predominantly poor and black with many elderly. Elliott and Pais's (2006) analysis of survey data shows that those with higher incomes were nearly twice as likely to have evacuated before rather than after the storm, and that low income blacks were most likely to have remained.

Of the people left behind, many died during the flood and its immediate aftermath. Sharkey (2007) provides the most detailed and careful analysis of the available fatality data, revealing a striking bias towards elderly people. As shown in Table 6.1, the death rate amongst the over-65s was 15 times higher than for the

Table 6.1 Age and race profile of Katrina's victims in New Orleans

	Number of victims	*Death rate per 10,000 people in the city*
All fatalities*	555	11.5
Under 65	184	4.3
White	31	2.7
African-American	151	5.0
Over 65	371	65.5
White	140	52.4
African-American	214	74.2

Source: Adapted from Sharkey (2007).
*The total fatality statistic used here is a subset limited to those for whom demographic data was available; the actual total is higher.

under-65s – 67 per cent of the flood victims were over 65, whilst this age group made up only 12 per cent of the city's population. Death rates were also higher amongst the black population (both elderly and non-elderly) than would be expected given the make up of the city, reflecting the greater severity and depth of flooding in largely black areas. Elderly people died not only because of drowning, but also because they couldn't access the medicines and treatment they needed (Fussel 2006) and because they were trapped in nursing homes, hospitals and their own homes:

> We have now learned how many elderly were left to die in nursing homes while they waited for transport – even though each nursing home was required by law to have transportation in place for evacuation ... Disaster relief coordinators were apparently unaware of the existence of frail and ill people trapped in nursing homes and hospitals, and focused efforts on transporting the generally healthier but more visible group taking refuge in the Super Dome. The numbers of elderly receiving in-home care that were left stranded in their homes will probably never be accurately known.
>
> (Foxman *et al.* 2006: 652)

Eventually, mass evacuation of the city was organised by various means, but not until thousands had endured living in appalling conditions for five or six days (including in the Superdome, Convention Centre and in hospitals and prisons), some for up to two weeks. Access to food became a particularly severe problem (Pyles *et al.* 2008). Some 96 per cent of New Orleans residents in the end left their homes and had to live elsewhere for a period of time. As summarised by Cutter (2006):

> Those that could muster the personal resources evacuated the city. With no welfare check (the hurricane struck near the end of the month), little food, and no help from the city, state, or federal officials, the poor were forced to ride out the storm in their homes or move to the shelters of last resort. This is the enduring face of Hurricane Katrina – poor, black, single mothers, young, and old – struggling just to survive; options limited by the ineffectiveness of preparedness and the inadequacy of response.

Relocation and return

On leaving the city, people found many different locations and forms of refuge. Some were able to return relatively quickly to damaged but still liveable properties, but these tended to be better off and white homeowners. Those who were poorer and previously renting their home were far more likely to be living away for an extended period (Sastry 2009) and were faced with rents which increased dramatically as demand outstripped supply. More wealthy evacuees were able to stay in hotels or apartments, whilst poorer ones were reliant on living in someone else's home or staying in trailer parks dispersed over a wide area.

The locations of the trailer parks set up by the Federal Emergency Management Agency (FEMA) to house longer term evacuees were themselves controversial. Davis and Bali (2008) found that these were resisted by some communities, such that race became a significant predictor of trailer park location, with predominantly black communities more likely to agree to host a trailer park than predominantly white. Several studies of local residents' attitudes to Katrina evacuees have found evidence of racial prejudice in perceptions of economic and social threat (Hunt *et al.* 2009; Shelton and Coleman 2009).

Health, economic and social impacts after the flood

Health impacts from the flood continued long after the immediate aftermath. The number of deaths recorded as being a result of Katrina increased over time. Of the 1,577 Louisiana residents who by June 2006 had died as a result of the disaster, 210 died after February 2006 and 480 died outside of the state following relocation (Sharkey 2007). Less severe health impacts were widespread, including injury, illness and long-term stress. Whilst some argued that 'there was a general sense that access to good health care was available for anyone that needed it' (Wilson 2006: 153), evidence suggested that through much of the early phase this was not the case. Although emergency provision was available, many of those affected were too poor to have health insurance, thereby limiting their longer term access to sufficient medical care. There were also acute shortages of hospital beds, equipment and staff reported long after the immediate disaster (Katz 2008). Subsequent to the immediate flood, many concerns were expressed about the presence of toxic contaminants in the flood water – a 'toxic gumbo' made up of oil, sewage, flooded chemical plants and waste tips, predominantly located near to minority populations – and the deposition of these contaminants into homes and onto soil across the city (Frickel and Vincent 2007). Despite official reports that played down levels of risk to health from these contaminants, community groups were deeply sceptical about the limited monitoring and analysis that had been undertaken and, by working with independent scientists, pointed to exceedences of safe levels, particularly of arsenic in the soil (Godsill *et al.* 2009). They also generated data on mould growth in severely damp homes and the health effects of exposure to mould spores, finding extremely high readings long after the water had receded. Generally a lack of confidence in official agencies undermined people's trust that the environment they had returned to was safe, contributing in turn to higher stress levels and anxiety.

Many lost their jobs as a result of the flood, with strong differentials by class, race and gender. A month after the flood, black workers were 3.8 times more likely to have lost their jobs than white workers, with this likelihood amongst black workers also differentiated by income (Elliott and Pais 2006). Women found it harder to re-establish employment that had often been part-time and low paid, particularly in the tourism sector. Education was severely disrupted for many children, because of both the destruction of schools in the city (Katz 2008) and the loss of experienced teachers. The impacts of relocation and living in temporary accommodation (Chauvin *et al.* 2008; Picou and Marshall 2007) also led to significant stress problems for children as well as adults (Hensley and Varela 2008), although the resilience shown by relocated children has also been noted in focus group research (Pfefferbaum *et al.* 2008). In various studies race was found to have a strong influence on stress levels (Elliott and Pais 2006), as did parental status and gender, women showing higher stress levels than men (Galea *et al.* 2008) due to the pressure of looking after children, the loss of social networks, and the home-making and hidden emotional work they undertake (Enarson 2006). Mental health services struggled to supply the scale and degree of support needed (Madrid and Grant 2008).

Recovery and reconstruction

There have been deeply contentious and extended debates about the reconstruction of New Orleans, founded on a history of deep political conflict over the development and governance of the city (Burns and Thomas 2008). Many alternative strategies and plans have been put forward, ranging from not rebuilding at all, given the vulnerable location of New Orleans as sea levels rise, through to rebuilding the city 'bigger and better' with more disaster-resilient structures. From an individual householder's perspective there is no doubt that some were much better able to repair, re-establish and re-occupy their homes than others. Many poorer residents had no flood insurance. Masozera *et al.* (2007) found a strong negative correlation between poverty and flood insurance cover in the city. Those with insurance and other access to resources, particularly where they lived in less damaged districts, were able to return relatively quickly and re-occupy their homes. This process was not at all straightforward, with insurance companies reported to have done all they could to delay and avoid paying out, but resources and capacities to act were available. Those without insurance, private renters and those who had lived in public housing found any return to the city much slower and more difficult (Katz 2008). Various obstacles also acted against the most needy being able to access grants and assistance to aid recovery (Bates and Green 2009; Morse 2008); for example, the federal 'Road Home' programme provided cash assistance only to homeowners whose houses were damaged by the storm, and not to renters. A new population profile for the city was therefore seen to be in danger of emerging, a profile that excluded many of those that had lived there before, changing voting patterns and therefore decision-making about reconstruction. Critics also complained that much of the reconstruction work was going to out-of-state corporations rather than local businesses, doing little to bring flows of employment and income to those most affected by the disaster and enabling 'profiteering' by major corporate concerns at the expense of the city's poorer residents (King 2009). Katz (2008: 24) characterises the 'market logic' of the government at all levels in critical terms:

At worst, they have pandered to corporate interests in 'theme parking', gentrifying, and whitening New Orleans, as can be seen in the padlocking of public housing and the too easy recourse to condemning and razing housing in damaged areas of poorer neighborhoods without giving former residents a chance – let alone any support – to rebuild their homes, at the same time as resources flow to reconstruct the tourist city of New Orleans.

After much local protest, community participatory processes have now been utilised in formulating neighbourhood plans for rebuilding the worst affected areas. However, concerns have been expressed about the lack of representation of those still displaced from the city, and the degree to which the involvement of lower income residents was really being achieved. Henry (2010: 237) concludes that the most likely outcome of long-term recovery will be continuity rather than social change, 'continuity in risk exposure, in inequality, in residential patterns, in ideological frames'.

Katrina and dimensions of injustice

This outline of research into Katrina demonstrates a deep intertwining of environmental risk and social difference. It has undoubtedly provided a compelling illustration of the forms of inequality in flood exposure and inequality in flood vulnerability discussed earlier in the chapter. But the discussion has gone much further in showing how the intertwining of flood risk and social difference has been produced through historical processes of urban change, economic and political marginality, deeply embedded discrimination, underinvestment in infrastructure and state institutions, and unevenly structured processes of preparedness, responses and recovery that served often only to exacerbate rather than in any way address differences in vulnerability. Centring on institutionalised racism (discussed in Chapters 3 and 4), Young (2006: 43) emphasises the long-term structural processes forming the background against which the disaster played out:

> Some commentators have tried to affix blame for a specific set of events ... In our anger we rightly assume that the devastation that Katrina wrought was (in part) preventable. In order to support that understanding, though, we have to look at governmental, economic and social processes for decades before Katrina – processes that were part of the background, assumed to be normal and acceptable. Focusing on assigning blame for a time-bound set of recent events and actions deflects from the longer structural view.

The flood experience was differentiated by race, but also by age, gender and class such that multiple elements of identity intersected. This differentiation also moved along the flood disaster cycle, manifesting itself in connected but distinct forms, from well-before to well-after the hurricane event. We have examined in this account two of the elements of the claim-framing framework introduced in Chapter 3 – *evidence* of multiple forms of inequality, and accounts of *processes* explaining how inequalities have been produced – but have not yet clearly pulled

Table 6.2 Concepts of justice in the Katrina flood

Justice concept	Explanation	Examples from Katrina
Distributive justice	the distribution of goods and bads	• flooded areas were largely poor and black • those who were unable to evacuate were disproportionately poor, black and elderly • those who died were disproportionately elderly, and black • those losing jobs were predominantly poor and black and women's jobs worse affected than men's
Procedural justice	access to information, participation in decision-making	• lack of involvement of vulnerable groups in preparedness and emergency planning • lack of information on toxicity risks from flood water • lack of involvement of affected communities in reconstruction
Justice as recognition	patterns of respect, stigmatisation, discrimination	• racist history of housing development in the city • lack of respect for lives of elderly and the immobile left to fend for themselves • racism in responses to trailer park locations

out normative concepts of justice and how these figure in the Katrina case. Justice will provide a more direct focus for the last section of the chapter, but in preparation and summary we can usefully consider the experience of Katrina in terms of the categories of distributive justice, procedural justice and justice as recognition. Table 6.2 tabulates these three forms of justice, drawing together examples from across the disaster cycle stages.

All three justice concepts are clearly implicated and also clearly intertwined – for example, the racially structured lack of recognition in past housing policy, producing uneven racial patterns of distribution of flood exposure, in part explaining who was included and excluded in decision-making over reconstruction. Katrina may not be entirely typical of flood experience across the US, or across the rest of the world, but it at least resonances in some dimensions with broader patterns. It also provides us with a rich foundation for considering, in a more systematic way, what justice in relation to flooding could or should constitute.

Justice and flooding

One of the obvious qualities of environmental justice compared to other ways of framing environmental concerns is the up-front prominence of justice. In the context of perspectives on flooding this is significant in shifting or adding to our view of the potential or actual flood victim. Whilst it is vital to reveal differences in vulnerability – in who is most at risk from flood impacts and why – a downside of this can be to emphasise weakness and victim status. This is particularly the case where a concern for vulnerability is translated into measurements and indices (Birkmann 2006; King 2001), and the role for management then becomes one of managing existing vulnerability in various ways. The danger here is that vulnerability and victimhood become a given, rather than something that needs to be more fundamentally addressed. One strength of an environmental justice framing is that, whilst focusing on inequalities in vulnerability, it can also readily make claims and assertions about what constitutes justice and fairness for people at risk of flooding. People and communities at risk become not only vulnerable but also citizens with rights to be asserted, achieved and protected.

Actually determining what justice in relation to flooding should consist of, what its metrics should be, is not straightforward (Johnson *et al.* 2007). Some 'big questions' (Cigler 2007) are raised about responsibility, about the roles of individuals, governments and private markets in creating flood disasters, and about how far the state should have responsibility for ensuring public safety and enabling or directing recovery. After Katrina many in the US were deeply critical of the performance of government agencies eroded by funding cutbacks, and asserted that far more should have been done by the state before, during and after the flood (Young 2006). Some others, in stark contrast, argued for much less state involvement and that leaving decision-making to private-sector actors and the incentive structures of the market would do a much better job:

> No national guard or other representatives of the state should be brought in ... Instead private police agencies, appointed by property owners, should deal with the looters. Further no tax money should be poured into New Orleans. These are stolen funds and should be returned to their rightful owners, the taxpayers of the nation. Of course this applies, in spades, to those victimized by Katrina. But the refunds should be in the form of money to specific taxpayers, not generalized expenditures for rebuilding, which their property owners may or may not favour. Instead, New Orleans should rely on private charity. Private enterprise should alone determine if the Big Easy is worth saving or not ... Possibly a Donald Trump type might try to buy up all the buildings at a fraction of their previous value and save his new investment by levee building and water pumping.
>
> (Block 2006: 238)

Leaving such extreme libertarian positions aside, we can work through a number of potential determinations of and elements to justice in the context of flooding in distributive and procedural terms. The intention here is *not* to arrive at a single preferred, universally applicable determination, but rather to lay out a number of alternatives, to which others could certainly be added (Johnson *et al.* 2007), and to explore each of their possibilities, strengths and weaknesses.

Equality of exposure, likelihood or impacts

As we have seen with other topics considered in this book, the notion of distributive justice as a *simple equality* in which exposure to flooding is somehow equalised such that all people, or all social groups, equally share in this exposure, or are equally represented on flood plains, is pretty nonsensical. Whilst disasters are man-made, floods do have a degree of natural agency; they happen in some places rather than others. They are not intentionally 'spreadable' over space, unlike, say, waste sites, in order to achieve fair shares across the population. Somehow manipulating flood-plain population profiles so that they work out equally at some scale of analysis is similarly nonsensical as a justice target. There are also more fundamental difficulties. The very equal pattern of aggregate population exposure to river flood risk across deprivation bands in England (shown in Figure 6.2) somehow appears to be 'a good thing' – all people regardless of deprivation being statistically, in aggregate, equally represented on the flood plain. However, this does not necessarily represent a just situation. For example, there may be very different degrees of choice involved across these deprivation bands – some choosing to buy a house in a riverside location, others having to locate to a flood-prone area in search of affordable rents – suggesting that the process by which the status of living at risk was arrived at may need to be part of the justice metric. Also, as we have seen, there can be very different vulnerabilities and coping capacities between those who are wealthy and those who are poor – aggregate equality of flood exposure does not therefore mean aggregate equality of flood impact.

A second alternative could be to conceive distributive justice in terms of a *maximum level of likelihood or probability of being flooded*, and to allow no population to be exposed to a level of risk greater than this. All in society (regardless of class, age, race, gender) would therefore be protected to a common likelihood of being flooded. This has some similarity to the setting of the air quality standard discussed in Chapter 5, which also seeks to achieve an equal level of protection. In principle this could be achieved through modelling and mapping flood scenarios and probability levels, strictly controlling where new building is allowed on the flood plain, installing physical protection in the form of flood defences for existing populations to reduce risks, and potentially removing existing populations from high risk areas. Whilst these measures describe some elements of current flood management practice, for example, in the UK, again there are conceptual and practical problems with conceiving justice in these terms. First, enormous faith is placed in the capacity for flood probabilities to be reliably calculated, something that is routinely undermined by unexpected and unusual flood experience and also complicated by ongoing climate change. Second, as shown in countless examples including that of Katrina, flood defences can fail or be overtopped. Third, as before, an equal level of flood likelihood does not account for unequal vulnerabilities to the impacts of floods. Fourth, controlling occupation of the flood plain can be very problematic, particularly where there is pressure for development for social reasons, where planning rules are weak, and/or where people are forced onto marginal flood-prone land in order to establish or sustain their livelihood. Whether justice in a wider sense can be achieved by clearing informal settlements and/or taking apart established communities is open to question. Mustafa (2005: 583) emphasises that much rests on *how* relocation is carried out, stressing the need for 'culturally appropriate and locationally convenient' replacement housing and the close involvement of citizen groups throughout. See also Dixon and Ramutsindela (2006) for an environmental justice framing of urban resettlement processes related to flood risks in South Africa.

Given that both of the above options fail to focus on differences in how floods are experienced, a better formulation might be to aim towards *equalising flood impacts*. Whilst this does appear to better recognise differentials in vulnerability, a key problem is how, when and at what scale impacts are to be measured. As we have seen, the impacts of flooding are very heterogeneous (death, stress, loss of employment, diminished education, property damage and so on) and can extend over long time scales, so that capturing all of these in some aggregate measure is not straightforward. Conventionally, in the case of flooding (and other fields) impacts are calculated in terms of monetary costs, and decision-making over, for example, investment in flood defences is typically carried out through varieties of monetised cost-benefit calculation. Relying on monetary values can, however, have perverse effects, such as the damaged homes of richer people being valued as far more important that those of poorer ones, or the lives of high-income earners being valued far more highly than those of low-income earners or of elderly people at the end of their income-earning potential. Whilst modified valuation techniques can be used to try to avoid such outcomes, and using non-monetary metrics might also be possible, a singular focus on impacts raises more difficulties than it resolves. Equalising impacts can also appear to imply a satisfaction with a certain level of impact rather than a desire to reduce this wherever possible (with parallels to the reduction of waste rather than its sharing out – see Chapter 4), although a notional equalisation level could be set to drive down losses over time.

Equalising the capability to be resilient

Given that these ways of thinking about justice are variously problematic, another direction to move in is to focus less on the agency of the flood (its occurrence, likelihood and impacts) and more on the agency of those potentially and actually experiencing flood events. By focusing justice on concepts of coping-capacity or resilience, a formulation of justice is arrived at which centres far more on Sen's notion of capability as the 'preferred space' for determining what is just (see discussion in Chapter 3). The capability to achieve valued functionings – such as health and well-being, livelihood and living with respect – is diminished by the impact of flooding. Focusing justice on achieving and sustaining these basic capabilities through the extended period of flood disasters – or, in other words, on the distribution of being resilient to the impacts of flooding – opens up justice to the entire flood cycle and to the many different processes which make people more or less vulnerable, or more or less resilient. Seeing justice in these terms would mean:

- Effective pre-flood preparedness would be ensured for all forms of community, with awareness programmes and emergency plans drawn up that were responsive and sensitive to differences in language, culture, gender, age and capability to act.
- Warning and evacuation processes would be designed in ways that ensured that all people were able to receive warnings (including those without access to communication technologies), and capable of achieving speedy and effective evacuation if this was needed, some needing more assistance than others.
- Emergency services and other agencies would treat people without discrimination, ensuring that all had places to relocate to that were affordable and safe to live in, and with access to health care.
- Assistance would be given to help people sustain incomes and re-establish employment, and resources would be available to all (for example, through universal access to flood insurance) to enable them to repair damage and re-establish their homes in a more flood-proof condition, or to move elsewhere if they so wished.

This is a demanding set of conditions, but they encompass much of what has been called for in the more normative disasters literature, including in post-Katrina analysis, for example, the need for culturally appropriate risk communication (James *et al.* 2007; West and Orr 2007), gender-fair emergency relief (Enarson 2006), effective insurance cover for both households and communities (Burby 2006) and a whole set of post-disaster rights as captured by the National Urban League's 'Katrina Bill of Rights' (see Box 6.3).

Box 6.3 The Katrina Bill of Rights

The Right to Recover – immediate help to get people back on their feet and rebuild their lives, including through extended unemployment assistance and a victims' compensation fund.
The Right to Vote – ensuring that evacuees have full voting rights in their home states, so that they have the voice they want and deserve in the rebuilding of communities.
The Right to Return – every evacuee must be guaranteed the right to return to their home, a three-year federal tax holiday for lower income returnees, and the right of first refusal to reclaim property.

The Right to Rebuild – every resident has the right to rebuild and have a say in what the future of their home will be. Rebuilding must take place in a manner that doesn't benefit only the big contractors and real estate developers. People should be protected from impacts on their credit ratings, from predatory lenders and from foreclosure proceedings against property.

The Right to Work – every resident must be assured of the right to work, with reconstruction jobs going to local people and businesses, ensuring that fair wages are paid. Longer term we must push for an economy that will sustain good-paying jobs for people of the region.

Source: Summarised from Morial (2009: xvi–xviii).

Such demands all push towards state intervention that targets public resources at the most vulnerable, a policy approach that Johnson *et al.* (2007) argue is in accordance with various specifications of justice principles.

Participatory justice and recognition

The list of ways in which specific elements of a 'capability to be resilient' might be achieved implies far more than simply action by the state. NGOs, community groups, local businesses and residents themselves are generally also recognised as playing key roles in resilience building. This brings us to questions of participation and recognition and the formulation of flood justice in these terms. Broadening the profile of those involved in all aspects of hazard and disaster management has been widely called for. As in other contexts, the rationales are varied and centre on the normative democratic principle that 'people at risk have a right to participate in decision-making about those risks', the calling to account and scrutiny of scientific expertise (Godsill *et al.* 2009), the broadening of perspectives and bringing of local knowledge into planning processes, and the making of better-grounded and socially respected strategies and decisions. Enarson (2006) captures many of these rationales, linking participation with recognition in her call for meaningful involvement specifically by women in the post-Katrina period:

> But we have learned that the most urgent need of all is for those most affected to reclaim their sense of place, some degree of control and autonomy, and the certain knowledge that their views count too in the re-imagining of the future. Will women's voices be heard in the independent commission likely to be appointed to review the national response to Hurricane Katrina? Will community recovery meetings be held at times convenient to those with children and in places safe for women? ... Measures are needed now to ensure women's representation on all public bodies making recommendations and decisions about the use of private and public

relief monies. Those women most hard-hit by Katrina must take the lead and men and other women must learn to listen. Women must be heard speaking out (and disagreeing) as elected officials, technical experts, community advocates, health and human service professionals, faith-based leaders, tenant association members, workers and employers, environmental justice activists, daughters, mothers and grandmothers.

There are tensions, as in other contexts, around to what extent reliance can be placed upon participation to achieve just outcomes and the scales at which participatory processes should operate. In the flooding and disasters literature most attention has been given to community-scale processes in developing preparedness and emergency plans and sometimes in formulating strategies for local flood management (Osti 2004; Pearce 2003; Speller 2005; Young 1998), matching with the scale of demands typically made from a justice perspective. However, involvement in higher level strategic decision-making, including in the allocation of resources, is arguably also needed, particularly in the context of adaptation to climate change impacts (see Chapter 8). There is also a danger that, in promoting participation in flood mitigation, responsibility is passed over, leading to an individualisation of risk in which people are made responsible for protecting themselves, regardless of the differential resources and capabilities they may have for achieving this (Brown and Damery 2002; Steinführer *et al.* 2007). Maintaining an inclusive and *collective* sense of process and participation therefore has to be a necessary part of procedural justice in this context.

Summary

Disasters provide moments in which fractures and inequalities in society are exposed for all to see. In the case of Katrina, we saw how the shocking playing out of the impacts of flood water in a city in the richest country in the world revealed countless ways in which the disaster discriminated by poverty, race, age and gender. Who was flooded, who died, who suffered health impacts, who was unable to recover and rebuild reflected deeply embedded processes that produced vulnerability across the disaster cycle. Whilst Katrina is the pre-eminent disaster to have been analysed through an environmental justice frame, the focus on inequality and discrimination connects it to a much longer standing literature on the unequal production of vulnerability to flooding around the world. Recent research has provided evidence of the socio-spatial patterning of who is living with flood risk in other places and at wider scales, producing diverse results that reflect methodological choices in analysis as well as the particular context within which urbanisation processes have operated along rivers and coastlines.

There is also much evidence that suggests that whilst the experience of Katrina victims through warning, flooding, evacuation and recovery may be extreme in the developed world, it is certainly not unique, and that social difference is a feature of many flood disaster experiences. Inequalities work out in different ways, and where there is a stronger state and social welfare framework the degree to which vulnerabilities are recognised and governmental responsibilities are acted upon can make a significant difference. However, in much of the developing world, where disasters interact with discrimination and grinding structural poverty, injustice can be readily apparent and framing flood disasters as a matter of justice can add edge to demands for better and fairer responses. There are various ways of conceiving justice in relation to flooding, and the multidimensional and dynamic character of flood disasters arguably demands an approach – such as the capability perspective – that can encompass notions of distributive justice, procedure and recognition across the flood disaster cycle.

Further reading

There is an enormous body of work on Katrina. Bullard and Wright (2009) provide many resources within an environmental justice framing, whilst Daniels *et al.* (2006) is another wide-ranging edited book on Katrina. The Natural Hazards Centre at the University of Colorado has an extensive website of resources with links to government reports, books and journal papers: http://www.colorado.edu/hazards/library/katrina.html. The Social Science Research Council also has an excellent and provocative set of papers on 'Understanding Katrina' at its website http://understandingkatrina.ssrc.org/. Key recent papers that offer a more developed theoretical analysis include Katz (2008) and Freudenburg *et al.* (2009).

For flooding and vulnerability more widely, a useful set of presentations is available at http://www.geography.lancs.ac.uk/envjustice/eiseminars/index.htm, and the Flood Hazard Research Centre at Middlesex University in the UK has an extensive catalogue of research and publications. For flooding and inequality in the UK see Walker and Burningham (2011) and the reports available at http://www.geography.lancs.ac.uk/envjustice/downloads.htm. For the Hull flood case study see http://www.lec.lancs.ac.uk/cswm/Hull Floods Project/HFP_home.php.

7   Urban greenspace

Distributing an environmental good

Greenspace is maybe not as immediately compelling an environmental justice theme as waste, air pollution, climate change and other environmental concerns covered in this book. It is certainly one of the least explored in the environmental justice literature and has not been a main focus of activist campaigning. Greenspace therefore represents one of the newer topics to be positioned within an environmental justice frame. The claim-making that surrounds its presence, absence and contribution to quality of life, health and well-being also has some distinctive characteristics. For environmental justice activism, greenspace falls into the recent broadening of scope to include environmental goods rather than only bads, and the move towards more proactive rather than oppositional community work related to sustainability themes (Agyeman and Evans 2003). For researchers, greenspace has become both another topic for classic socio-spatial distributional analysis (who lives near to greenspace and who doesn't), as well as a focus for health, planning, leisure and urban design communities wanting to establish the contribution that greenspace provision could or should be making to healthy living for poor and marginalised communities (Floyd and Johnson 2002; Frumkin 2001).

As will become apparent, one of the challenges of working with the category 'greenspace' is that it can take many different forms, serve multiple functions and have multiple meanings. Whilst I try to take on some of this complexity in what follows, to make the task manageable the chapter focuses specifically on greenspace in urban areas. This limiting of scope reflects where most of the research literature has been focused, and also where the most acute lack of and need for greenspace is to be found. Living in a rural environment does not inherently mean that accessible, usable and personally productive greenspace will be available in abundance for all. Environmental justice is also not inherently an urban issue, as there can be environmental inequality and injustice within rural contexts (see Agyeman 1987; Moore and Pastahia 2007). However, the pressure for development and maximisation of land values means that greenspace can be all but excluded from built-up environments, particularly where population densities are high and the public value of parks, gardens, urban woodland and other forms of greenspace has not been recognised in land use designations and planning schemes (where these exist). Urban greenspace can be scarce and intensely pressurised, available and accessible to some but not to others.

In terms of the cross-cutting themes of the book (see Box 7.1), this chapter will demonstrate again how questions of inequality and justice need to be carefully thought through, and how they have different parameters as we move from one domain to another. One of the themes brought alive by the case of greenspace concerns the implications that follow from dealing with the multiple values and meanings that can be given to an environmental feature. It is therefore with the meanings and use values of greenspace as a 'good thing' that we will begin.

Box 7.1 Greenspace and cross-cutting themes

By discussing greenspace as a feature of urban environmental inequality, we will be able to explore:

- the case for a pluralistic view of environmental justice and how this relates to complexities in defining 'what is to be distributed' when environmental features, such as greenspace, have malleable and multiple meanings and use values;
- the related methodological complexities involved in generating evidence of patterns of inequality;
- the relevance of need, and the uneven distribution of need across populations, when examining inequalities and injustice in relation to an environmental good;
- the role of history and geographical context in explaining the processes that have produced contemporary patterns of inequality in the availability of urban greenspace;
- the tensions and contextualities that can emerge when making judgements about the injustice of a situation and therefore the objectives to be sought after, for example in relying on procedural justice to produce just outcomes, or in viewing compensation through greenspace provision as a form of restorative justice.

Greenspace as a 'good thing'

It may seem unnecessary to spend any significant time discussing whether and why greenspace can make a positive contribution to people's quality of life. You or I might take that as read. However, if we are to make claims, to argue for the establishment, protection and promotion of urban greenspaces, to resist alternative land uses and call for the spending of public resources, it is absolutely necessary to have some well-founded arguments and evidence in place as to exactly why greenspace matters. As we shall see, why it matters also relates to the form(s) it needs to take and the qualities it needs to have. Furthermore, to simply assume that greenspace is a good thing may also mean to be blind to cultural and social difference. Others who are not from the same backgrounds as you or I may make quite different normative evaluations as to the importance and value of greenspace and the meaning it has in everyday life. Some evidence that that is the case will be discussed in later sections.

There are various disciplinary perspectives that have attempted to tackle the question of the personal and social value of greenspace, drawing on related work on the value of contact with nature, the importance of exercise and relaxation for health and the community, and the political role of public space. There are various excellent reviews of this growing body of literature (see further reading at the end of this chapter). Drawing from these reviews, five dimensions particularly relevant to an environmental justice framing emerge. These are summarised here without detailed reference to specific studies.

First, the links between psychological health and contact with nature have been increasingly documented, with explanations focused on the reductions in stress that can be achieved and the ability to focus attention away from the complex demands of everyday living. The notion of restorative environments has been used to convey the sense in which certain types of places can act in these ways to help people cope with stressful and involved lives, with depression and bereavement, mental fatigue, recovery from illness and so on. Whilst some of these claims relate to contact with nature more generally (including with animals and pets), or the benefits of being in countryside and wilderness, the significance of the urban setting is also emphasised. Urban greenspaces can act as oases of peace, tranquillity and calm close to where people live or work. They can provide relief from traffic and from crowded, noisy and sometimes confrontational living spaces. These effects may be particularly important for certain sub-groups, such as children, women and people with mental health problems (Frumkin 2001; Strife and Downey 2009). Just being able to see plants, trees, river corridors or open greenspace has been claimed in a range of studies to produce significant benefits in reducing stress and improving psychological well-being. Unrelieved stress has been shown to suppress immune responses and increase vulnerability to illness, and plays a part in a range of serious health conditions such as heart disease. Measures that can help to reduce stress are therefore of wider significance to improving health and addressing health inequalities.

Second, greenspaces can provide an important setting for physical exercise of various forms – walking, jogging, cycling, informal games, organised sports activities. Exercise has been increasingly seen as part of maintaining good health through different stages of life – including younger and older age – and particularly significant in addressing rising levels of obesity. Regular physical activity can contribute to the prevention of more than twenty conditions, including coronary heart disease, diabetes, certain types of cancer and mental ill health. Worsening patterns of obesity across many countries are of particular concern. The notion of 'obesogenic environments' (Lake *et al.* 2010) has been used to capture how many aspects of contemporary living environments have structurally contributed to the problem. Lack of greenspace for physical activity, alongside the deficiencies of mobility infrastructures of various forms, are identified as key elements.

Third, urban greenspaces, it is argued, can have community and social value, providing a public forum where neighbours can get to know each other and a space for collective recreational, cultural and social activity and for political debate and organisation. In these ways greenspace can help to build more cohesive neighbourhoods, develop social capital and support more active citizenship. Claims have been made for the ways in which activities based in greenspaces can help promote contact and integration between different age groups and racial and ethnic groups. For individuals, becoming more involved in community life can reduce social isolation, which in turn can reduce their vulnerability to stress, depression and general ill health.

Fourth, the role of greenspace in enabling 'food justice' has emerged as a distinct concern (Food and Fairness Inquiry 2010), connecting into environmental justice framing and activism in a number of countries and contexts (Gottlieb and Fisher 1996; Heynen 2006b). Food justice is about addressing access to healthy and affordable food for poor and marginalised communities and, in Global North contexts, has largely focused on the distribution of food stores of

Figure 7.1 An urban community garden providing for social space and food production in New York.
Source: The author.

different forms (for example, Frank *et al*. 2006) and the occurrence of what have been called 'food deserts' (Wrigley 2002). However, one of the advocated responses to food justice problems is to promote the growing of local food through either household or community farm initiatives, so that fresh, healthy produce is locally available. To enable this to happen, the availability of land is obviously important. Here greenspace takes on a quite different function and value, one that is about small-scale agricultural production, turning over available land to the growing of fruit and vegetables or the keeping of livestock. The contribution to health and well-being is through the consumption of healthy food, as well as other claimed productive outcomes for well-being related to economic productivity, community empowerment and self-determination, and the building of strong neighbourhoods.

Finally, greenspace can play a role in mitigating or moderating other forms of environmental risk or harm, including those covered elsewhere in this book. The greening of urban areas can play a moderating effect on air quality, with trees and vegetation acting as filters for particulates and absorbing some forms of gaseous pollution (Bolund and Hunhammar 1999). The greening of the urban environment can also play a role in reducing the risk of urban flooding, which can be lower where there is plenty of urban vegetation to intercept and absorb storm water. In terms of climate change, trees and dense vegetation act as a carbon sink, contributing to carbon mitigation. Greenspace can help reduce heatwave risks by providing shade and spaces for cooling off, and vegetation cover of roofs and walls can limit building overheating and the use of air conditioning (Alexandri and Jones 2008).

Always a good thing? Contested meanings of urban greenspace

In combination, these reasons why urban greenspace can be seen as a 'good thing' are substantial. Greenspace emerges as something of importance and value about which judgements of equality and justice might therefore be sensibly made. The evidence base is, however, not entirely uncontroversial or unproblematic, a recent review, for example, noting that whilst 'there is clear evidence for a positive relationship between greenspace and health', there is a lack of understanding of the mechanisms: 'the studies find associations, correlations or linkages but no cause and effect relationships. Moreover, the evidence comes from a range of types of data and indicators, some much less objective than others' (Bell *et al*. 2008: 61). Specific studies have been undertaken using certain methodologies, in particular places and under particular conditions, limiting their generalisability. But beyond the need to be evaluative of evidence claims, there are further reasons why we should approach greenspace, as an environmental justice issue, with some care.

First, urban greenspace is not always and only a good thing. Rather, it has the capacity to have a very different status and meaning, to move from good to bad in the way that it is viewed and evaluated (Burgess *et al*. 1988). A number of

Figure 7.2 Urban greenspace in north London.
    Source: The author.

studies have shown that some forms of greenspace, woodland in particular, can be associated with crime, with risks to safety, assault and violence (Fisher and Nasar 1992; Herzog and Chernick 2000; Jorgensen *et al.* 2002). This is both in generic terms – woodland being generally seen as unsafe and a place where 'bad things' can take place – and in relation to specific spaces where crimes have been known to happen or where certain activities or groups of people perceived as threatening are expected to be found (Madge 1997). As Byrne and Wolch (2009: 748) note, people visit parks 'for illicit reasons – from the prosaic to the potentially dangerous – including homelessness, voyeurism, exhibitionism, sexual gratification, drug use, thievery'. The status of greenspace can also change over time, so that at dusk or at night it can be viewed quite differently from during the day, or in summer quite differently from at other times of the year.

Second, and very much connected, there can be important cultural, gender, age and other differences in how particular forms of greenspace are viewed and evaluated. For example, women can have greater concerns about threats to safety in parks or woodland spaces than men (Burgess 1988; Jorgensen *et al.* 2007). Brownlow (2006a, 2006b), in a study of the Fairmount Park system of Philadelphia, shows how over time women (African-Americans in particular) and children have become excluded from this significant urban greenspace and, as a consequence, misrecognised in justice terms. He traces how changes in how the park system was managed, and how social control was exercised, led to increasing fear of crime and associations with neglect and decay. The park became a space of disorder and fear, rather than one of enjoyment and relaxation. Other studies have also shown that ethnic or cultural groups can perceive greenspaces differently and look for them to provide certain use values rather than others (Byrne and Wolch 2009; Floyd and Johnson 2002; Loukaitou-Sideris 1995; Rishbeth and Finney 2006; West 1989). Urban greenspaces have been seen to act as markers of segregation, forming barriers or walls between urban areas with distinct racial or economic profiles (Gobster 1999; Solecki and Welch 1995), spaces therefore in which social networks are truncated rather than strengthened. Byrne and Wolch (2009), in an extensive review, show how parks in some parts of the US have a history of being managed as racially segregated spaces – formally or informally – in which discrimination and misrecognition are practised and experienced. This is part of a whole set of spatially and historically constituted elements which they combine in a complex model of the relations between space, race and park use (see Figure 7.3).

Third, it is clear that the multiple values given to greenspaces mean that conflict and contestation may well be a feature of how they function in practice.

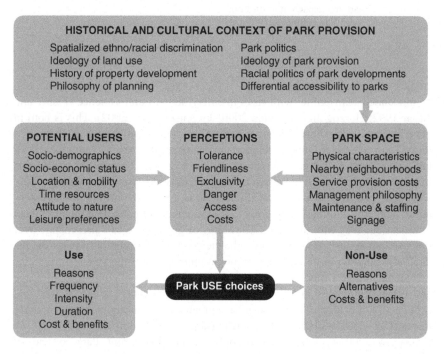

Figure 7.3 A model of the elements shaping park use.
   Source: Adapted from Byrne and Wolch (2009).

Different people and population groups can be looking for different uses and values – one person's successful social-networking readily constituting another's aggravation or threat. There can be conflict between, for example, those that are looking for peace and relaxation and those wanting to pursue noisy sport, social or recreational activities. Any one greenspace is unlikely to be able to simultaneously realise all of the positive values listed earlier, and quality, layout and design may all be important in how it is valued. Heynen (2003) makes wider points about the tensions in how urban greenspace is understood in tracing its multiple values across spatial scales. Focusing specifically on urban forests, he argues that locally orientated reforestation projects that seek to provide a more equitable and just distribution across a city may not be productive in justice terms when considered at other scales. In terms of contributing to global concerns about biodiversity and climate change, it makes more sense, he argues, to focus on ecological efficiency by building up and sustaining existing forest islands. As he points out, 'this plan would of course lead to planting more trees near upper/middle class areas, thus contributing even more to inequitable and unjust distribution of trees at a local level' (ibid.: 993). Thus there are scalar ramifications for environmental justice when moving focus from a stand of trees to a forest island, which 'can alter the merits by which we deem something to be unjust' (ibid.: 982)

In these ways, greenspace emerges as something rather more complex than we might have at first imagined. In terms of a focus on environment and social difference, it also shows much potential for producing, and playing into, patterns of inequality across multiple forms of social and demographic distinction – age, race, class, gender, disability and so on. It is therefore to the core question of inequality and how this can be assessed and evaluated that we now turn.

Greenspace and social difference: evidence claims and inequality

The assessment of patterns of greenspace inequality can be approached in different ways. Predominantly the environmental justice literature has assessed patterns of difference and inequality through some form of spatial analysis. The focus is on the distribution of the greenspace itself, where it is, the characteristics and qualities it has, and how this relates to patterns in the proximate population characteristics. Most of the following discussion will focus on this form of analysis.

It is important to recognise though that there is an alternative approach which focuses on the *users* of greenspace – how there are differences in the profiles of who makes use of and gets value from different forms of greenspace. There is a substantial and longer-standing literature taking this approach, deploying surveys and/or focus groups and interviews, that is in part linked to the health-related work already discussed but is also emerging from research fields focused on urban recreation, leisure science, regeneration and the management of urban space. This work highlights how there is 'under-representation' of particular population groups in their use of greenspaces (a lower proportion than their profile in the general population) and trends of change typically moving in less equal directions. Generalisation is problematic, but in a UK context, for example, research tends to show that older people are under-represented as greenspace users both in comparison with other age groups and in proportion to their presence in the residential areas surrounding the space. Ethnic minorities, women and 12–19 year olds are also underrepresented (Comedia and Demos 1995; Lucas *et al.* 2004; Urban Green Spaces Taskforce 2002). A range of 'barriers' or deterrents are typically identified as explaining these patterns, including poor accessibility, lack of seating and other facilities, safety and risks, vandalism and graffiti, and racial issues. As an example, a more detailed account of research focused specifically on children is provided in Box 7.2.

Box 7.2 Children and greenspaces

Children arguably have a particularly strong need for good-quality public greenspaces, but they can experience distinct inequalities in being able to make use of them (Day and Wager 2010). Children can get significant value from the exercise, social interaction, skills development and creative play that take place in greenspaces (Strife and Downey 2009). Taylor *et al.* (1998) describes them as 'spaces to grow'. Furthermore, there are wide-spread concerns about children spending longer indoors, being inactive, 'de-natured', and not getting the exercise they need to maintain fitness and a healthy body weight. Whilst these changes have multiple causes, lack of access to greenspace and recreational facilities are implicated, as are various constraints on these being utilised, constraints which can be experienced very unequally. For example, in focus group and interview research with children living in three locations in Scotland, Day and Wager (2010) found that outdoor public space was 'tremendously important' to all the children, but that for children living in urban deprived areas their use of such spaces was accompanied by a routine level of fear, worry about safety, territorialism and stigmatisation by adults. Children in less urban areas had a much richer environmental experience, with everyday access to prized green and natural spaces and much more scope for safe and creative play. These findings are echoed in research focused on low-income and minority communities in the US (Lopez and Hynes 2006). However, Strife and Downey (2009) argue that far more needs to be known both about how inequalities in access to, and use of, greenspace are experienced by children and why these inequalities exist, identifying this as a significant gap in environmental justice research.

Environmental justice work involving spatial analysis can be seen as focusing on a particular, and fundamental, form of 'barrier'. In examining patterns of where greenspace is, and where it isn't, differences in the opportunities for people to use or get value from greenspace are revealed. If there is no local greenspace, then no local value can be obtained. On the other hand, we need to remember that just because local urban greenspace is physically present and spatially proximate doesn't necessarily mean that people *will* use it, will *want* to use it, and get value from it. Some forms of use value, such as building shading and pollution filtering, that are about background physical processes may act independently of human engagement, but most others do not. We will return to such questions later in the chapter, but they should also be kept in mind through the rest of this section as various forms of statistical spatial analysis are examined.

Some methodological complexities

It was stressed in Chapter 3 that there are methodological complexities in how any analysis of socio-spatial patterning is undertaken. Greenspace is no different, but there are some distinctive complexities which merit specific examination.

Most fundamental is the question: what counts as greenspace? We have already touched on the varieties of urban greenspace that might be relevant, and Table 7.1 provides a categorisation which further demonstrates this variety. This poses some questions for anyone undertaking spatial analysis. For example, should 'private gardens and grounds' (Table 7.1, row 2) be included as a form of urban greenspace? They might be very important for day to day encounters with nature, for relaxation and some forms of physical activity, but as private spaces with restricted access for others should they be included? How about the indicated types of 'amenity greenspaces' (Table 7.1, row 3) such as 'road verges' – should they be included? And how big does a 'green' piece of ground need to be to become 'greenspace'? Where is the threshold between a small patch of grass and something more significant?

Determining these questions fundamentally has to interact with the purpose of the study and which of the multiple values that greenspace can have is of interest. If the focus is on opportunities for exercise, then only certain spaces (of certain sizes) where exercise is feasible would be included. If the focus is on opportunities for growing food, then a different set of spaces would be relevant (in Table 7.1, row 8, allotments would be included, but churchyards and cemeteries presumably not?). If the focus is on the health benefits of relaxation, then another sub-set would be involved – and so on.

These questions in turn have direct relevance for the mechanics of the analysis. If the focus is on the visual benefits of greenspace, then maybe the general presence of density of greenspace and nature in an area is what matters – trees, flower beds, gardens, public and private spaces that make a visual impact would all be relevant. If, however, it is about relaxing within a park or green corridor, then having access is crucial and some form of distance from where people live becomes significant – ideally not just a crow-flies distance but something that

Table 7.1 Types of open space

Type	Description
Public parks and gardens	Areas of land normally enclosed, designed, constructed, managed and maintained as a public park or garden.
Private gardens or grounds	Areas of land normally enclosed and associated with a house or institution and reserved for private use.
Amenity greenspace	Landscaped areas providing visual amenity or separating different buildings or land uses for environmental, visual or safety reasons, i.e. road verges or greenspace in business parks, and used for a variety of informal or social activities such as sunbathing, picnics or kickabouts.
Play space for children and teenagers	Areas providing safe and accessible opportunities for children's play, usually linked to housing areas.
Sports areas	Large and generally flat areas of grassland or specially designed surfaces, used primarily for designated sports, i.e. playing fields, golf courses, tennis courts, bowling greens; areas which are generally bookable.
Green corridors	Routes including canals, river corridors and old railway lines, linking different areas within a town or city as part of a designated and managed network and used for walking, cycling or horse riding, or linking towns and cities to their surrounding countryside or country parks; these may link green spaces together.
Natural/semi-natural greenspaces	Areas of undeveloped or previously developed land with residual natural habitats or which have been planted or colonised by vegetation and wildlife, including woodland and wetland areas.
Other functional greenspaces	Allotments, churchyards and cemeteries.
Civic space	Squares, streets and waterfront promenades, predominantly of hard landscaping that provide a focus for pedestrian activity and make connections for people and for wildlife, where trees and planting are included.

Source: Scottish Executive (2003).

represents a viable walking or maybe cycling route (you might live 200m from a park but need to walk 800m to get into it). The population groups of interest are also important, and it is clear from the discussion so far that many social differentiations are relevant to greenspace issues. A focus on children, for example, would bring forward certain values and uses, and therefore certain greenspaces into the analysis, whereas a focus on older people might bring forward quite different ones.

These and other more generic considerations (scale, comparison populations, statistical methods and so on) make quite a few demands on the process of research design. They also make major demands on data (Commission for Architecture and the Built Environment 2009). Ideally, a wonderfully precise and complete data set on greenspaces would exist, with plenty of information about the type of greenspace, access conditions and routes, facilities, quality of upkeep, etc. The reality can be quite different, necessarily constraining both the scope and to some degree the logic and sense of what can be accomplished.

An example: greenspace inequalities in Scotland

My own experience of being involved in a study with a group of colleagues examining greenspace in Scotland is instructive (Fairburn *et al.* 2005). The research was funded by a consortium of organisations with varied interests in environmental justice issues, including some with specific interests in greenspace and woodland. The overall study covered a range of other environmental concerns (air quality, derelict land, industrial sites, landfills, quarries, river water quality), and the funders were interested in how patterns varied with deprivation and it was intended to be national in scope. This made examining greenspace problematic. Most of Scotland is rural, much also relatively wild and uncultivated, meaning that a general analysis of inequalities in greenspace across the country would do little more than show up obvious urban–rural differences; if you live in rural Scotland, you have more nearby greenspace than you would in urban Scotland. No surprises there, then. Moreover there was no national spatial data set on greenspaces in general, from which we might be able to extract particular types, or just those in urban areas. Thus a full national analysis was unfeasible on a number of grounds.

There was, however, a national-scale dataset specifically for woodland. Using this, we were able to conduct a reasonably meaningful national analysis, but still

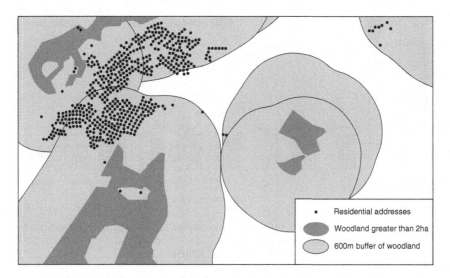

Figure 7.4 Example of 600 metre 'buffers' around woodland in Scotland.
Source: Adapted from Fairburn *et al.* (2005).

with plenty of limitations. We decided the focus would be on the use of wood-lands for recreation and exercise and, given that there are general rights of access to land in Scotland, it was reasonable to presume that all woodlands could *potentially* be used in this way – although actual access conditions and usability would in reality be very varied. A distance of 600 metres from the outline of each wooded area over two hectares in size was used for analysing population proximity. The figure of 600 metres came out of a review of literature on distances people were prepared to walk to nearby greenspaces, but it could only be applied as a straight-line distance (as in Figure 7.4 on previous page). Anything more compli-cated was impossible given the scale of the study and the resources available.

In reporting the results of the study, we were able to distinguish between urban and rural populations (using a definition of urban areas generally applied by the government), and also between established woodlands and new woodlands planted since 1995. The results showed that people living in the most deprived areas were less likely to be near to woodlands when considering all woodlands (although the differences were not that striking), but that for new woodlands (Figure 7.5) there was a bias towards more deprived populations (rural and urban), reflecting, we were told, policies that had been followed for several years of giving grants for woodland that could be accessed by deprived populations. Thus, the analysis was completed and evidence was produced, but with many caveats necessarily surrounding the claims that could be made and the implications that could be derived.

To work with a broader definition of greenspace, it was clear that we would have to shift to a sub-national scale. At the time only one city in Scotland,

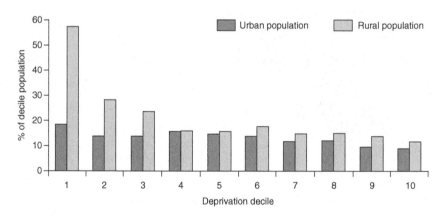

Figure 7.5 Percentage of the Scottish population within 600 metres of new woodland by rural and urban deprivation decile.
       Source: Adapted from Fairburn *et al.* (2005).
Note:
1 Decile 1 value for rural population is high but based on a very small rural population overall in decile 1.

Glasgow, had suitably differentiated land-use data available to enable this. This big data set identified 'outdoor recreation' as a land-use category, which predominantly took the form of greenspace but also included sports stadia and the like. Polygons of outdoor recreation land were available, but no other information on their more specific type, quality or access conditions. An analysis of populations living within 300 metres (chosen as a distance that children would be prepared to walk) of the identified land parcels was undertaken in two different ways, one using a straight-line distance and the other a far more involved network analysis that followed 300m routes along walkable pathways. Perhaps counter to standard expectations, both methods showed that the *most* deprived populations in Glasgow were *more* likely to live within 300m of greenspace (see Table 7.2, which compares the 10 per cent most deprived with the 10 per cent least deprived). The difference between population groups shifted quite significantly depending on the method of distance analysis used, the route access method showing greater proportional differences.

Making sense of these results highlighted the limitations of the analysis. The greater proximity to greenspace for the most deprived groups in Glasgow can be accounted for by the movement of much of the poorer population of the city, through slum clearance, to the outskirts of the city in the post-war years. Large council estates were sited specifically to give access to green fields and cleaner air. These have now become peripheral estates with major social problems, and whilst there is evidently much greenspace around blocks of flats and between paths and roadways, the quality of this space and its ability to provide *any* significant positive use value is very open to question. 'Purposeless green desert' was the description given by one of the project funders. It was also noted that some of the greenspace captured in the analysis was derelict land rather than managed spaces (not that derelict land is entirely without informal recreational value).

In summary, evidence of the social patterning of forms of greenspace in parts of Scotland was produced, but only in limited and particular ways. Resource and data constraints, combined with the complexities involved in translating the nuances of multiple forms of use value into an empirical analysis that made logical sense, clearly shaped and constrained the evidence that was produced.

Table 7.2 Proximity to greenspace for the top and bottom 10 per cent of income data zones in Glasgow

Income deprivation	Percentage within 300m of greenspace – straight line method	Percentage within 300m of greenspace – route access method
Top 10% most deprived people	92	62.3
Bottom 10% least deprived people	75	34.8
All Glasgow	79	44.2

Source: Adapted from Fairburn *et al.* (2005).

Studies from North America: parks, activity spaces and trees

Other researchers have grappled with such methodological problems more effectively than we were able to at the time. Most empirical work to date has been undertaken in North America, and the discussion below reviews five studies from the US and Canada which demonstrate various methodological approaches and derived outcomes.

A number of US studies have focused specifically on parks, as managed and typically very multifunctional public greenspaces. Wolch *et al.* (2002) analysed 224 park spaces across Los Angeles (LA), asking which areas of the city are 'park-rich' and which are 'park-poor'. Their focus was on the health implications for young people, but differentiated in terms of race, ethnicity and socio-economic status. Using density measures (park acres per 1,000 residents) and a distance of quarter of a mile from the park edge to indicate accessibility, their analysis identified a pattern of provision starkly differentiated by income and race (p. 20):

> Considering park acres per 1,000 residents, it is clear that low income and concentrated poverty areas have relatively low levels of park resources and access. Moreover, African American, Asian-Pacific Islander, and Latino dominated neighborhoods, where almost 750,000 children live, have extraordinarily low rates (1–2 acres per 1,000 total population, and 3–6 acres per 1,000 children) compared to white dominated areas (with almost 17 and 100 acres per 1,000 total population and children, respectively) where only 235,000 children reside. In those neighborhoods that are almost exclusively Latino, the number of total park acres, and accessible park acres per 1,000, are shockingly low: less than an acre per 1,000 population in park acreage, and less than 2 acres per 1,000 population living within a quarter mile of existing parks.

These strong differences are partly explained by white dominated areas being located on the edge of the urban area, and close to very large regional parks. But they also reflect the racially fragmented development of Los Angeles, and the way that at different times varying amounts of open space and parkland were set aside for public purposes. Historical context again is crucial. Wolch *et al.* (2002) also examine the allocation of funding for the improvement of park and recreation spaces, concluding that this was being unevenly distributed and poorly targeted. Neighbourhoods with the largest shares of young people received half as much funding (on a per youth basis) than areas with the lowest concentration of youth, and richer areas were receiving as much if not more funding than poorer ones. Joassart-Marcelli (2010) also finds wide disparities in funding for parks and recreational spaces across the wider LA region. Municipalities with large minority populations and low-income ex-urban communities typically have the lowest expenditure, a pattern which is linked to a wider problem of changes in the funding of park provision, including a greater reliance on community fund raising (discussed further below).

Maroko *et al.* (2009) examine parks and physical activity spaces in New York City. Their findings are far more equivocal than those for LA, and they emphasise the methodological complexities and interpretative difficulties involved. Using two forms of analysis, they find what they call 'unpatterned inequality', with parks and physical activity sites unevenly distributed across the city in spatial terms; however, this distribution is not associated overall with race/ethnicity, income or any other examined variables. Through more intensive case-study work focused on two parks they then discuss the complexities involved in making a detailed and rigorous assessment that 'better reflects reality' (ibid.: 21). As they conclude:

> Looking at one factor at a time is likely to result in misleading findings. Therefore, a more complex model that accounts for as many different types of variables as possible (park size, access points, barriers, network distance, perception of safety, crime rates, park maintenance, availability and variation of physical activity sites) will be needed to develop a more accurate measurement of park accessibility.
>
> (ibid.: 21)

Gilliland *et al.* (2006) take a slightly different approach again by specifically studying the recreational opportunities available to children and youth. They examine publicly provided recreation spaces in London, Ontario (Canada), and map and analyse these 537 spaces in relation to neighbourhood characteristics (socio-economic and environmental variables, an index of neighbourhood social distress, and a neighbourhood play-space needs index). They conclude that there is no systematic socio-environmental inequity with respect to the provision of neighbourhood recreation spaces, but there are areas in the city where younger people have no access to formal play spaces. This leads them to argue for a targeting of future resources towards these areas.

Lindsey *et al.* (2010) focus on a different form of 'linear' greenspace – urban greenways or trails – in Indianapolis, Indiana, and similarly find no systematic bias against marginalised groups. Rather, they show that minorities and the poor have disproportionately greater access (defined as living within half a mile) to greenways:

> Although only 21 percent of the city-county population is African-American, 35 percent of the population in census tracts within one-half mile of these greenway trails is African-American. About 16 percent of the trail population is living in poverty, while only 12 percent of the city-county population is living in poverty.
>
> (ibid.: 339)

They explain this not as a function of deliberate progressive targeting of greenway provision on poor and minority areas, but as a historical outcome of the legacy of the natural landscape of the city and development processes that extend back more than a century:

> The city planned to create parkways along the streams that flowed to White River near the center city and partially developed them. The middle and upper classes that first occupied these fashionable areas then began moving out of them in the middle of the twentieth century in search of more spacious neighborhoods. During the 1960s and 1970s, movement of whites into outlying townships and counties accelerated, and African Americans moved into the once-fashionable neighborhoods along the parkways. Now, thirty years later, these populations are the beneficiaries of decisions to develop the greenways system.
>
> (ibid.: 343)

Finally, a different indicator of urban greenspace, or maybe greening more generally, is examined by Landry and Chakraborty (2009) in a study of the distribution of public right of way or street trees in Tampa, Florida. These trees are publicly financed, raising issues of equity in the allocation of resources as well as in patterns of tree location. Through using a high-resolution data set to map tree cover and associate this with patterns of race and ethnicity, income and housing tenure, they find that residents of neighbourhoods characterised by low-income households, renters and African-American populations have significantly fewer street trees and therefore lack the same access to the benefits provided by these trees as areas populated by more affluent, white homeowners.

Looking across this sample of a growing but still relatively small body of studies – producing quite different conclusions on evidence of inequality – we could ask the question: so is urban greenspace unevenly distributed across social groups? And the answer would be a very resounding 'well, it depends'. It depends on where you are talking about, what type of greenspace is examined, what contribution this is expected to make to well-being, what forms of social difference are analysed, what historical patterns of urban development have been involved in the places you are looking at, and so on. If we look beyond the US to studies undertaken in other parts of the world – for example, Timpiero *et al.* (2007) on public open space in Melbourne, Australia, and Pedlowski *et al.* (2002) on urban forests in Rio de Janeiro, Brazil – such an equivocal response becomes even more appropriate.

A 'well, it depends' answer may feel a bit unsatisfactory, but it does not diminish the value of generating evidence through socio-spatial analyses. These do not *only* have value if they all point in the same direction towards some universal systematic process being at work. It is good to be able to identify that in some places greenspace does appear to be more progressively distributed and that people are (hopefully) benefiting as a result (the glass is half full); just as it is important to identify where this *isn't* the case and where demands for action might be made. This begs the question though of the basis on which that judgement and demand might be made, to which we now turn.

Greenspace and justice

Despite the difficulties involved in making sense of evidence and methods, we can still step back and ask what the end target in justice terms should, or could, be. What constitutes an environmentally just situation for urban greenspace? Is a revealed socio-spatial inequality sufficient to sustain a claim of injustice, or are other parameters and modes of reasoning necessarily involved? As for other environmental topics, there are various possible ways of answering this question, each of which brings forward different understandings of what justice constitutes.

As in the case of air quality, aiming towards a simple equality, a situation in which everyone in cities should somehow have an equally positive experience of greenspace (no more and no less), is inherently problematic. This implies some form of uniform urbanism that has so many trees per square kilometre (no more and no less), so many square metres per square mile or per person of green area for exercising, so many minutes per day of use of that greenspace, or some other similarly problematic measure. To understand justice in such terms would be to deny the agency of nature and landscape within urban areas – the linear greenspace of river corridors, for example, cannot exist in equal measure across a city or between cities – and also the differences between people in how they might value and want to make use of greenspace (as discussed earlier).

At the other end of the spectrum, it is also possible to talk of justice and greenspace in very libertarian terms, to say that it is all a matter of choice. Here the idea is that people decide where to live and weigh up the value of living near to greenspace against other positives of residential locations, and that people also decide what use to make of greenspace against other potential uses of their time and resources. Under this rationale no further judgement about justice, independent of the outcome of choices as freely exercised, is necessary. This position is also deeply problematic, as it denies all sorts of limits on choice and the roles of powerful actors (property developers, planning and city authorities, wealthy homeowners) who determine what is and has been historically possible, and how public and private greenspace is to be invested in, or not invested in. See Heynen (2003, 2006a) for excellent accounts of how such processes have operated historically in Indianapolis.

Minimum standards and the uneven distribution of need

What does begin to make more sense and does typically provide the basis for claims made by activists and advocates for greenspace provision is to talk about justice in terms of the meeting of minimum standards – such that greenspace in effect becomes a form of 'right' of urban living. Under this reasoning, people can live with varying amounts of greenspace, but everyone should have some minimum level of access or provision. This approach has been argued for within both academic and policy communities. For example, in England a greenspace access standard was adopted by English Nature in 1996 which specified that there should be accessible greenspace provision of at least 2 hectares per 1,000 people and that no person should live more than 300m from their nearest area of greenspace (English Nature 1996). When setting a minimum standard, as an expression of a just situation, there is important detail in exactly how the standard is defined and the criteria it includes (as also discussed in Chapter 5 in relation to air quality standards). For example, if the standard is a distance from home, there would be details to address in terms of how this is measured and what constitutes the greenspace that has to be within the minimum distance. Does this greenspace have to be of a minimum size, meet a quality standard and/or be capable of being used for certain purposes? Ideally, any minimum standard would therefore include more than just proximity, but the more complex the standard the smaller the degree of flexibility in fitting this with varied real-world contexts, and the harder it becomes to measure, monitor and achieve.

A further consideration though is that a common minimum standard may not be targeting attention at where greenspace is most *needed*. When discussing justice in relation to air pollution and flooding, the need to take account of differing vulnerabilities to harm was examined. The flip side, when looking not at risks but at environmental goods, is that taking account of the need for the good may be necessary to achieve a just situation. It could, for example, be readily argued that poorer communities have a greater need for local public greenspace than richer ones, and therefore should have a higher minimum standard (or be prioritised in meeting a common standard). Poorer households have fewer resources to finance travelling to greenspace at a greater distance, or in general to escape the pressures and stress of city living through trips and holidays. They are less able to afford other forms of exercise – gym membership, classes, joining a cycling club. They typically live in greater population densities, with more noise, less privacy and quiet, and fewer or smaller private gardens. They also typically have worse health profiles, so that the health benefits of local greenspace could have a greater impact. Whilst such generalisations may not always apply, we can see how a justice claim could be constructed here not for a common minimum standard but for an approach that differentiated this standard from area to area, or which targeted investment and action in poorer communities. Other forms of targeting, for example, by age – see the work by Wolch *et al.* (2002) and Gilliland *et al.* (2006) discussed earlier – could also be argued for in order to achieve justice in the distribution of greenspace that reflected a distribution of need.

Both rights and need-based arguments have been deployed by environmental justice groups campaigning for greenspace provision within their communities. For example, in New York the establishment of a water-front park on the edge of Harlem was campaigned for by WE ACT for Environmental Justice as a site of escape, relaxation, socialising and relief from noise and heat stress for a poor, mixed-ethnicity area of the city. This high quality, multi-functional, public greenspace (see Figure 7.6) replaced a marginal strip of land which had been blighted by industry, dereliction, crime and drug use, but it was only put in place through a hard-fought battle with the city authorities, who were investing in greenspace along other parts of the waterfront in downtown Manhattan but not in Harlem

Figure 7.6 Harlem Piers water-front park in New York.
          Source: The author.

(West Harlem Environmental Action 2004). A similar set of arguments were also made in the Bronx, where the advocacy work of Sustainable South Bronx led to the city investing over three million dollars in a water-front park at Hunters Point, one of the poorest neighbourhoods in the city, with high unemployment, poor health and many social problems. Local community participation showed that more parks were wanted and needed by local people, as long as they were 'safe, accessible and well maintained' (Sustainable South Bronx 2008). Further initiatives being taken forward by the group under its 'Green the Ghetto' campaign include a South Bronx Greenway of bike and pedestrian paths and a neighbourhood tree planting programme.

Greenspace as compensation

These cases from New York could also be interpreted as a form of compensation in which justice is achieved by redress for past and/or existing wrongs. Both the Bronx and Harlem have been the location of environmentally damaging and polluting activities such as sewage works, waste transfer stations and bus and lorry depots. In earlier and ongoing phases of activism, both WE ACT and Sustainable South Bronx have mounted environmental justice campaigns against these activities, arguing that poor minority communities were disproportionately suffering from the waste and pollution flows that were sustaining rich, downtown Manhattan. A park in this context can be seen as a form of community-scale compensation for having lived with a concentration of pollution – a balancing of a bad with a good. Whether this constitutes an adequate interpretation of 'doing environmental justice' is contentious. In the US the approach used by the Environmental Protection Agency (EPA) in implementing its environmental justice policy has, in part at least, been in effect one of compensating communities for hosting polluting activities. In a careful analysis, Holifield (2004) shows how much of the response of the EPA to identified 'environmental justice communities' took the form of community development programmes and employment and local greening initiatives, rather than action focused directly on the waste or industrial activity of concern. Whilst such actions may provide some restitution for past and present impacts, trading a park for pollution may equally be seen as a 'sop', an attempt to placate and smooth over local antagonisms rather than deal with the fundamentals of the problem and the patterns of inequality that exist. Sweeping generalisations are problematic though. Much rests on the detail of each case, as the sequencing and combination of actions, and the views and interpretations of the actors and communities involved, can be crucial in making sense of why particular investments in parks have been made.

Procedural justice and participation in greenspace planning

This brings us to another perspective, which is to frame justice and greenspace as a matter of participation and procedural justice. There are many grounds for involving local communities in the process of planning, investing in, establishing and maintaining public greenspaces. There are substantive arguments that through participation greenspaces will be better designed, better reflect local needs and values, better address the needs of multiple users and also tap into local creativity and enthusiasms. There are also instrumental arguments that by involving the community there is less likely to be resistance to interventions and investments by public authorities, less conflict between users and better longer term care for the space if it is felt to be locally 'owned' and part of community identity. Normative arguments go further to say that people have a democratic right to be involved in greenspace decision-making and planning processes, and that such involvement should be inclusive, recognising cultural and social difference, and implemented in ways that meet all the conditions of a procedurally just process (see discussion in Chapter 3).

Such perspectives, as in other situations we have considered in this book, can be integrated with notions of justice focused on distribution. So a claim for distributional justice reflecting need – which, for example, targets investment in low-income areas – can be readily combined with one that argues that the process of investing in greenspace needs to be a participatory one that directly involves local people. Exactly along these lines, the establishment of the Harlem Piers park was preceded by a community visioning process, so that 'community voices were heard from the outset' rather than 'outsiders designing, refining, configuring, contorting and controlling our dream' (West Harlem Environmental Action 2004: 15) But again there are balances to be struck and judgements to be exercised about how far a reliance on community participation should go and what form it should take. Perkins (2010), in an analysis of processes working in the US (but arguably applicable elsewhere), traces how a decline in public funding for parks has led to increasing use of 'shared governance' arrangements in which corporations, not-for-profit organisations and local residents increasingly become involved in and take on responsibility for park management, upkeep and maintenance. Whilst this sounds like a positive step towards more democratic processes of decision-making in how local resources should be allocated, in practice, he argues, it can mean a greater reliance on market-based solutions which service some interests and exclude others. 'Communities of self-interest' can develop around parks in some locations, whereas in others without direct state investment they fall into disrepair and neglect or fail to be established anew. As with neoliberal moves more generally that have passed functions that were once the responsibility of the government onto non-state actors (Heynen *et al.* 2007), more marginal communities can be burdened by the need to volunteer and become involved in shared governance arrangements, rather than empowered by the possibilities these open up. As Perkins (2010: 264) concludes, it is necessary to formulate ways that 'assist all disadvantaged communities in democratically participating in greenspace governance if we decide that communities of self-interest are in fact superior to previous forms of state provision'. As noted in other contexts, when it comes to procedural justice, much rests on the detail of how 'community involvement' and 'public participation' are realised, and how the wider political system that local processes fit within is configured.

Summary

In this chapter we have examined urban greenspace as an example of an environmental 'good' that has recently come within an environmental justice frame, and that we might expect to be distributed quite unevenly. Five varied reasons why availability, access to and use of greenspace have been claimed to be a good thing were identified – psychological health, physical health, community value, growing healthy food and mitigation of environmental risks – all of which imply the need for certain qualities of greenspace to be achieved. It was, however, stressed that urban greenspaces can also be seen as places of threat, insecurity and risk. Some people may therefore get many positives out of a greenspace experience which others are, or feel, entirely excluded from, with differences by age, income and race highlighted in the literature. Different use values may also conflict or be in tension, as they can work in different ways and at different spatial scales – from the very local value of being able to exercise in a park to the global value of urban trees contributing to climate change mitigation.

These multiple uses and meanings of greenspace have implications for the generation of evidence of inequalities, which, within an environmental justice framing, has tended to focus on spatial patterns in the local availability of green-space, i.e. who has it and who hasn't. We saw how there are methodological complexities in designing and undertaking such studies and varied outcomes from the limited body of research that has been undertaken. Some studies, of some forms of greenspace in some places, have found inequalities by age, income and/or race, whilst others have not. This highlights both the significance of spatial and historical context in shaping how greenspaces have become included or excluded from urban areas and also the ways in which different study designs can generate different outcomes. Finally, as in other chapters, we have worked through a number of ways in which justice in greenspace provision might be conceived. Focusing on the setting of minimum standards of accessibility makes much sense in distributional terms, but there are complexities involved in defining standards and there is a strong case for taking account of differences in the need for local provision. It is also possible to see the provision of greenspace as a form of compensation for taking the burden of the forms of environmental disamenity discussed in other chapters. Focusing on local participation in green-space decision-making and governance provides an additional way of thinking about justice in procedural rather than distributional terms.

Further reading

For reviews of the literature on the relationship between greenspace, health and well-being see Bell *et al.* (2008), Frumkin (2001) and Tzoulas and Korpela (2007). Also see the special issue of the *Journal of Environmental Psychology* on 'restorative environments' (2003, vol. 33). Strife and Downey (2009) provide an excellent wide-ranging discussion focused specifically on children. The journal *Health and Place* has many recent articles relevant to the chapter themes. For a much more pithy attempt to capture greenspace health benefits as a 'Vitamin G' see Groenewegen *et al.* (2006). If you are interested in following up further on environmental justice and access to food see Gottlieb and Fisher (1996) for an early discussion, Heynen (2006b) for an urban political ecology perspective and the Food and Fairness Inquiry (2010) for a broad global perspective.

Byrne and Wolch (2009) provide a wide-ranging review of geographic research on parks and race, also developing an agenda for future research. Maroko *et al.* (2009) include a useful tabulation and review of previous empirical studies on inequality and parks as part of their New York study. Heynen's work on urban nature, and forests specifically (Heynen 2003, 2006a), expertly brings together environmental justice and urban political ecology perspectives and merits a detailed and careful reading. More can be found out about the work of WE ACT and Sustainable South Bronx from their websites at www.weact.org and www.ssbx.org.

8 Climate justice

Scaling the politics of the future

Compared to every other concern discussed in this book, climate change makes the most persuasive case for a justice framing. With climate change we are confronted with evidence of patterns of inequality and claims of environmental injustice that span the globe, that permeate daily life and which pose threats to the current and future health and well-being of some of the poorest and most vulnerable people around the world. Climate change demands more than ever that we think relationally, about how things interconnect, about who benefits at the expense of others and about the spatially and temporally distant impacts of patterns of consumption and production. The consequence is that, for many already economically, politically and environmentally marginalised people, climate change presents compounding forms of injustice.

In Chapter 2 we saw how environmental justice had globalised vertically to move beyond the local, regional and national scale to encompass environmental concerns that operated internationally. Climate change is the exemplary case. Environmental justice groups in the US, and elsewhere, have added climate change to their established campaigning themes and set up specific new programmes. These include the 'Environmental Justice and Climate Change Initiative', based in the US, which describes itself as 'a movement from the grass-roots to realize solutions to our climate and energy problems that ensure the right of all people to live, work, play, and pray in safe, healthy, and clean environments' (Environmental Justice and Climate Change Initiative 2008). This initiative aims to bring together environmental justice, climate justice, religious and policy networks to promote 'just and meaningful climate policy' through leadership training and advocacy work. In Canada 'Just Earth' describes itself as a 'coalition for environmental justice' focused on carbon mitigation and similarly puts forward profiles of actions for individuals and organisations and a declaration calling for the setting of ambitious targets and commitments by the Canadian government. Such activity has interfaced with a surge of discursive framing and mobilisation around 'climate justice', described by Pettit (2004) as a 'new social movement for atmospheric rights'. Climate justice framing hasn't necessarily drawn on environmental justice traditions but shares much in common with them in being concerned with questions of difference, inequality and fairness (Angus 2009). Climate justice has provided the organising frame for regional coalitions

Figure 8.1 Climate justice march in London.
         Source: Michael Gwyther-Jones.

such as the Pan African Climate Justice Alliance, for groups traditionally focused
on development, such as Oxfam, and for religious groups such as CAFOD.

Partly because of this campaigning work, justice has figured very directly in the
debates that have gone on around the major international meetings attempting to
negotiate global agreements on emission reductions and resource commitments.
Box 8.1 shows just a few examples. In these debates some very different positions
have been taken about what justice or fairness means and how important these
principles are compared to questions of effectiveness, efficiency and urgency in
climate policy. A key objective of climate justice campaigning has therefore been
to assert that justice does and should matter across all forms of policy and intervention concerned with climate change adaptation and mitigation, and at all scales.
As we shall see, climate justice can be conceived across different 'units of analysis' or communities of justice; hence national and local responses raise fairness
and inequality issues just as international negotiations between countries do.

In this chapter I will explore the very multidimensional character of climate
justice, considering how both patterns of impact and vulnerability and patterns
of responsibility are structured along lines of social difference of various forms.
I shall show how different concepts of justice – distribution, procedure and recognition – are interwoven across these dimensions, and that these in turn can be
considered at different spatial scales and across generations (looking both back-
wards and forwards from the present day). However, before moving into more of
the detail of climate justice, it may be useful to lay out some of the overarching

Box 8.1 Examples of how justice figures in positions on climate change negotiations

'We demand an equitable international negotiation process that acknowledges, respects and advances the concerns of vulnerable communities everywhere, both in the Global South, and in the United States. We call for a real, accountable, and just transition from fossil fuel dependency to a more localized green economy.'

(Extracts from a letter from the Movement Generation Justice and Ecology Project to President Obama, 2009: http://www.movementgeneration.org/ dear-president-obama)

'This is not a perfect agreement, and no country would get everything that it wants. There are those developing countries that want aid with no strings attached, and who think that the most advanced nations should pay a higher price. And there are those advanced nations who think that developing countries cannot absorb this assistance, or that the world's fastest-growing emitters should bear a greater share of the burden.'

(President Obama, prepared remarks for the Copenhagen Climate Change Summit, 18 December 2009)

'It has become clear that the Danish presidency – in the most undemocratic fashion – is advancing the interests of the developed countries at the expense of the balance of obligations between developed and developing countries.'

(Interview with G77-China chief negotiator Lumumba Di-Aping at the Copenhagen Summit, 14 December 2010)

'Pressure is mounting on developing countries. Those who suffer the injustice of climate change are encouraged to be "constructive", while those who caused it "divide and rule" through new country categories (e.g. "most vulnerable"), or offers of early – but profoundly inadequate – finance, or other means. We, the undersigned people and organizations of Africa, call for a fairer and more science-based solution to climate change. We, as Africans, stand ready to play our part. But cooperation must be based on justice.'

(African Climate Justice Manifesto, Pan African Climate Justice Alliance http://www.pacja.org/)

Box 8.2 Climate justice and cross-cutting themes

By discussing climate change through an environmental justice frame we
will be able to explore:

- the importance of understanding the distribution of vulnerability to
  harm and the distribution of responsibility for harm and how these can
  interrelate to create multiple forms of injustice;
- the significance of spatial and temporal scale in defining patterns
  of distributional inequality, shaping how evidence is collected and
  analysed, how generalisations across space and time are made and how
  policy decisions are contested;
- the challenges involved in generating evidence of uncertain patterns of
  current and future inequality at global through to local scales;
- the three core forms of (in)justice – distribution, procedure and
  recognition – and how these can all be found in an analysis of climate
  justice debates;
- the challenges involved in specifying and achieving procedural justice,
  particularly in the context of a multi-scale, uncertain and intergenerational problem;
- the need to take account of the 'side effects' of environmental policy
  in assessing patterns of inequality and claims of injustice.

---

challenges and elements of the activist, academic and policy debate. Box 8.2
summarises the cross-cutting analytical themes highlighted in this chapter.

Challenges for climate justice

The bringing of justice into the centre of the highest profile contemporary
'environmental' concern presents many challenges. Climate change and justice
are interwoven in involved ways, with the complications identified by early
academic work (for example, Waterstone 1985) being largely amplified rather
than resolved by later analysis (Adger *et al.* 2006; Beckman and Page 2008;
Ikeme 2003). Three challenges concerned with matters of knowledge, scale and
power are crucial.

---

1 My position is in the mainstream in that I am persuaded to take the threat of climate change
   seriously, not only by what I have read of the scientific case and the processes of assessment
   being used by the Intergovernmental Panel on Climate Change (IPCC); but also by the enormous risks involved in denying the scientific consensus that human-induced climate change
   is a real phenomenon, and in due course being proved catastrophically wrong. My view, in
   the face of uncertainty, is therefore guided by degrees of both trust and precaution.

Knowledge claims and uncertainty

Knowledge claims about climate change as a *physical* process are intensely contested and disputes between climate change advocates and sceptics have repeatedly dominated public debate.[1] However, the uncertainty that is inherent to climate change science is also characteristic of knowledge claims about patterns of inequality in climate change impacts, adaptation and mitigation. Uncertainties in the evidence of the distributional impacts of climate change stem from many sources – the predictive, forward-looking work involved, the detailed spatial differentiation that is hard to model with any precision, and the systemic socio-environmental-technical interactions which will lead to unknowable and unexpected outcomes. The research base is also as yet limited in its extent; Adger *et al.* (2006: 1) comment that 'research on equity in the environmental arena ... fails to address the multiscale and multifaceted issues produced by climate change and its impacts'. We therefore need to remain very open to shifts and evolutions in our understanding of the inequalities of climate change and the working out of climate justice as we move forward into a warmer and (hopefully) more carbon-conscious world. What we now expect of the spatial and social differentiation of climate change may not be entirely what we observe or expect in future decades. As Beckman and Page (2008) comment:

> The problem, of course, is that each generation of policymakers is faced with a profound lack of knowledge of the future with the result that they lack more than a minimal basis from which to establish the long-term impacts of alternative environmental policies.
>
> (ibid.: 530)

Key questions therefore revolve around how political processes do or do not take up uncertain forms of evidence and the extent to which evidence of social distribution and inequality is 'on the table' to be considered, and viewed as reliable or providing 'sufficient knowledge' to be useful (Dow *et al.* 2006). One example of how there are both uncertainty and politics wrapped up in evidence related to issues of justice is that of greenhouse gas (GHG) emission data. GHG data does not come from direct monitoring of exactly what carbon, methane or nitrogen dioxide is being released from where (a truly impossible task!), but rather from protocols for producing estimates and modelling emission totals from proxy data, such as sales of petrol feeding into estimates of carbon emissions from transport (see Chapter 3 on proxies). This, from the beginning, introduces some element of uncertainty, as protocols have to include assumptions and generalisations, and choices have to be made about what emission sources to include and how. When aggregating the emissions from a country, there is then the question of what 'sources' to attach to that country. This is straightforward for, say, a power station located within national borders and using locally sourced fuel, but how about flights to and from a country or shipments of goods from overseas? Who 'owns' and has responsibility for these emissions? The point here is not that GHG emission data is flawed and cannot be trusted, but that there are different ways of making calculations and aggregations, which can shift the evidence base on which comparisons are made and thereby the strength of the claims about justice that are put forward by different interested parties. For these reasons part of the intense politics of climate change negotiations has been about exactly how evidence of GHG emissions is to be calculated, disaggregated, reported and utilised.

A second example, relating to climate change impacts rather than emission reductions, is that of extreme weather events – floods, typhoons, hurricanes, forest fires, droughts and the like. These, of course, happen sporadically in various parts of the world, in different forms and in variable frequencies and intensities. When they do, a potential evidence claim is to link the specific event to climate change, to make it part of a wider climate discourse and set of processes, and to build this evidence into claims about the injustice being suffered by those who are affected by climate change and the apportioning of blame for their suffering. Major floods, as discussed in Chapter 6, repeatedly generate this form of claim-making. This is uncertain scientific territory though, because any one event is hard to link definitively in causal terms to wider processes of climate change, and apparent changes in frequency or intensity can be due to fluctuations for reasons other than general global warming. Again then, the politics of evidence comes to the fore. For some, the need to bring attention to climate change and the unstable and unequal future we are heading into is sufficient to warrant making the linkage from event to wider process. For others who are more cautious, or who seek to deny both instabilities and inequalities, there is more than enough reason to challenge and dismiss such claims.

Scaling climate justice

A second related challenge is how we understand climate justice in scalar terms – both spatial scales and time scales. As briefly discussed in Chapter 3, scale is an important parameter of how justice concepts are applied and evidence claims are constructed. For climate change though, the choices made about scale are particularly significant and the range of possibilities particularly wide. Predominantly, climate justice has been examined at a global scale. In negotiations and campaigning and in academic analysis of what constitutes justice in climate change policy, the world space and its population have been divided up into nations and blocks of nations – Global North and South, rich and poor, developed and developing, G7 and G77. Here the detail of exactly how such blocks are made up, by whom and for what purpose, matters – note the reference to 'divide and rule' and 'new country categories' in the extract from the Pan African Climate Justice Alliance in Box 8.1. Such categories can reveal, to different degrees, differences in per capita greenhouse gas emissions and differences in the severity of expected impacts, such as on future food availability, livelihoods and vulnerability to hazards. They also have implications for how negotiation and decision processes are structured and interests are represented in these. Competing notions of fairness in global climate justice are therefore in part about competing ways of spatially dividing up the world.

We can also move away from a global analysis to look within national borders. Aggregating and averaging people over units of political space both reveals and hides, such that statistics like 'national per capita' can mask important intranational distributional differences in both emissions and impacts. Rich elite consumers in poorer countries can be seen as just as responsible for current GHG emissions as those in the Global North, and, as Hurricane Katrina demonstrated, vulnerabilities to environmental extremes can be found even in the seemingly best resourced of societies. There is therefore a rather tricky task to be performed of simultaneously holding together different scales of analysis in how climate justice is understood.

It is not though only about spatial scale; for climate justice, how time is viewed is also important. Concerns about changes in the climate and their consequences for human well-being are to a large part about the future, and about the interests of future generations. Climate policy therefore has to confront to what extent intergenerational justice and the needs of future societies are balanced with intragenerational justice and the needs of the present day (Beckerman and Pasek 2001). Different positions can be taken, and there have been intense debates about how well we can know what the needs of future generations will be, and also how people 'yet to be born' can be fairly represented in democratic decision processes (Beckman 2008; Beckman and Page 2008; O'Neill 2001; Thompson 2005). As noted earlier, there are also matters of history and past generations. Carbon and other GHG emissions have been made for a long time, rising steadily and then more rapidly as industrialisation and consumerism have advanced over the past two centuries. These emissions have accumulated in the atmosphere, contributing to current levels and to the warming that has already taken place. As will be examined more closely later in this chapter, for some actors climate justice has to take account of this carbon legacy, or 'ecological debt', adding past emissions to the responsibility accounts of the countries that industrialised earliest and for longest. For others the past is irrelevant, a responsibility that has passed rather than one to be factored into current formulations of what fair climate policy should constitute.

Power and 'lock-in'

For questions of both knowledge and scale it is clear that there is a politics to defining what and how we know and to constructing climate justice claims. This brings us to the third overarching challenge: the many structuring effects of power. In Chapter 3 questions of power were a central part of the discussion of processes through which environmental inequalities and injustices are produced, replicated and sustained. As has been laid out in an extensive academic literature, power comes in many different forms, some obvious, others far more subtle and hidden. Power is related not only to the work of obviously powerful actors (big governments and transnational corporations) but also to the ways that everyday life is structured by the form of cities and infrastructures, by technologies and cultural expectations, norms and routines that serve to 'lock in' established ways of doing things – inevitably to the advantage of some and the disadvantage of others.

For climate justice the many forms that power takes, and the many forms of inequality that result, are particularly significant. Carbon, in particular, is so deeply embedded in contemporary capitalist society – in the way that economies operate, accumulation is maintained, money is made and livelihoods are sustained – that processes of attempted 'decarbonisation' rub up against all sorts of vested interests and established hierarchies of wealth, influence and status. These interests and hierarchies can serve both to obstruct processes of change and, where they are taking place, to steer them towards protecting the status quo of economic and political power rather than towards any form of more progressive societal transformation and redistribution (Luke 2008). As Sayer (2009: 351) argues:

> The project of stopping global warming through regulation runs up against not only the accumulation based nature of capitalist economy and culture, but the enormous social-spatial inequalities it has generated, for they in themselves present a huge barrier to the development of collective responses.

Transition objectives are also confronted by the way that carbon-based energy consumption is locked into the layout and material form of cities and suburbs, and by the 'incumbent regimes' that dominate energy and transport systems. Globally it is also easy to be pessimistic about the prospects of generating significant new income flows or redistributions from Global North to Global South for adaptation or mitigation purposes, given the long history of uneven development and colonial and post-colonial practices that have dominated international relationships. Adaptation is also likely to be easier to imagine than to realise, with the costs, speed of change and complexities involved being rarely acknowledged (Adger and Barnett 2009). That is not to say that, for all these reasons, advocating climate justice is futile and that profiles of power and patterns of lock-in cannot change, but rather that the *process* of seeking climate justice has much to confront on a continual and ongoing basis.

Having introduced these three overarching challenges, we can now consider in more detail how justice in its different forms of distribution, participation and recognition is wrapped up in, first, climate change impacts and, second, climate change responsibilities.

Impacts, vulnerabilities and adaptation

Clearly, climate change is only a concern because of the nature of the impacts it is expected to have, the risks it will produce and the damaging consequences that will result for both current and future generations. Whilst part of the concern about climate change relates to its ecological consequences and impacts on species and ecosystems, the primary driving anxiety is the threat presented to future human well-being. Nothing less than total societal collapse, an apocalyptic end to all that we know, is predicted in the rhetoric of speech-makers such as

Tony Blair and Al Gore. In this mode of discourse, climate change becomes something universal, with consequences for all people in all places.

Whilst such apocalyptic representations serve a purpose in demanding public attention, they can also serve to hide the deep differences and inequalities in how climate change impacts are likely to be experienced and responded to. For Swyngedouw (2010) this is deliberate, a clear example of 'post-political' discourse in which universal threats and consensus concerns are used to evacuate contestation from public debate. Such moves make it all the more necessary to be absolutely clear about the variability of what climate change is likely to produce and lead to. A key concern of climate justice activism has been to do just that. In the discussion that follows, therefore, we move through evidence of patterns of differentiation in climate change impacts, first at an international scale and then within nations.

International patterns of differentiation

The variability in climate change impacts arises for three interconnected reasons. First, in physical terms global warming does not mean that everywhere simply becomes warmer (worth repeating for UK tabloid journalists, at least, who fail to grasp even this basic part of 'the science'). Impacts on weather patterns are expected to be very variable around the world, some regions and localities becoming warmer than others, some wetter, others drier, some with more weather extremes and so on (see the detailed regional reports of the IPCC for many examples). Exposure is therefore highly differentiated and not necessarily going to lead everywhere to adverse impacts; indeed, some are expected to be beneficial. Sea level rise as a consequence of climate change appears more constant, but even so will be experienced to different degrees because of simultaneous land mass movements that are already leading to shifts in land–sea relationships. The science is highly imperfect but the regional and local variability and dynamism of climatic processes are clear, meaning that heterogeneity in exposure is inevitable.

Second, some parts of the world, some forms of economic activity and some livelihoods are more closely related to and dependent on the climate and vulnerable to climatic change than others. Agricultural economies are particularly sensitive to changes in climate, for example, low-lying coastal areas are most affected by sea level rise, and water-scarce regions are more vulnerable to reductions in rainfall. Similarly, some people are more sensitive to the impacts of extreme weather events than others; for example, heatwaves largely lead to the deaths of older rather than younger people, in part because of their greater physiological sensitivity to higher temperatures. However, it is *crucial* to recognise that vulnerability is not simply a deterministic matter of location, physical characteristics and sensitivity. As made clear in Chapters 3 and 6, vulnerability is socially produced, part of the wider political economy and culture, which typically means that those who are already disadvantaged in material, cultural and/or political terms are also those who are most vulnerable to hazards of various forms, including those deriving from climatic variability.

This then introduces adaptive capacity as the crucial third element. The capacity and ability to cope with, and adapt to, changes in climate and their knock-on implications are far from constant. Some countries, economies, communities and households are much better positioned to adapt to climate change than others. Handmer *et al.* (1999) see this in global terms, identifying global adaptive trends which mean that most wealthy countries have 'substantial adaptive capacity' built upon robust institutions, access to insurance, access to knowledge and expertise and other resources. In contrast, they characterise parts of the developing world as being excluded from adaptive capacity because of 'social and economic processes, chronic violence, oppressive governments and warfare' (Handmer *et al.* 1999: 267).

Whether or not adaptive capacity can be characterised in such sweeping terms is open to debate. As Roberts (2009: 186) points out, 'people in the global South are not mere suffering victims, but have agency and autonomy, and are actively addressing this [climate] crisis every day'. However, that adaptive capacity is not held in equal measure is beyond question. For example, Tol *et al.* (2003), whilst recognising the overarching importance of poverty, strongly make the case for a differentiated view:

> What is also certain is that the poor will lose most, at least relative to their income. The poor are more vulnerable both because their exposure is higher and because their adaptive capacity is lower. If vulnerabilities are not uniform, one has to be very careful when aggregating impact estimates over countries or, for that matter, over economic sectors or income classes.
>
> (Tol *et al.* 2003: 25)

Adger *et al.* (2006) make the further point that strategies developed to promote adaptation can *themselves* produce winners and losers, depending on who is acting, the choices made and their knock-on consequences. They argue that leaving adaptation to the private decisions of individuals means that those with knowledge and resources are more likely to adapt in a timely and effective way than those without. Collective adaptation decisions, made locally or by national governments, may also serve to protect vested interests and reinforce vulnerabilities, including through the various processes of 'disaster capitalism' (Gunewardena *et al.* 2008).

The coming together of the three elements – variability in physical change, vulnerability and adaptive capacity – to produce an intense geographical diversity in climate change impacts (Yohe and Schlesinger 2002) is demonstrated in a wide range of regional assessments and national analyses. For example, Box 8.3 lays out the parameters of climate change impacts for Bangladesh, a country expected to be particularly vulnerable for various compounding reasons of physical and human geography, economic profile and high levels of existing poverty. Such expectations may not be fully realised (for example, histories of coping with extremes of flood and drought in Bangladesh may make the country more adaptive in some respects than others). However, as Dow *et al.* (2006: 91) argue, enough is known, based on experience to date with coping and not coping with hazards around the world, to enable 'ethically based initiatives on adaptation' to

Box 8.3 Projected climate change impacts in Bangladesh

When assessments of climate change impacts are made at a regional level, the degree of uncertainty involved multiplies, but some indication of likely consequences can still be derived. For Bangladesh, the assessments predict that the country will experience multiple physical changes – sea level rise, higher evapo-transpiration losses, stronger summer monsoons, more intense cyclones and connected coastal and inland floods, and potentially reduced dry season precipitation. Bangladesh is a very poor country with average income levels below those for the region. The population of 133 million (75 per cent of which is rural) is unevenly distributed, with very high population densities along the coast where most of the more significant physical effects from climate change would be experienced.

Figure 8.2 The complex coastline and fluvial geography of Bangladesh.

The country already experiences recurrent floods. In 1998 a severe flood displaced 30 million people and left 21 million homeless. It is estimated that 18 per cent of the land area and 11 per cent of the population would be physically displaced by a 1m sea level rise, with the compounding problem that options for within-country migration are limited. Sea level rise will also severely disrupt transport and water supply infrastructures. The total cost of sea level rise alone has been estimated at 10 per cent of GDP. The cost of constructing engineered defences to protect against sea level rise has been estimated at over one billion US dollars, far beyond what can be afforded. In terms of agricultural productivity, a 17 per cent decline in rice production and 61 per cent decline in wheat production are estimated, with fisheries also impacted by the combined effects of sea level rise and salt-water intrusion. The economy is significantly dependent on fishing and agriculture (making up 25 per cent of GDP). With declining production, population dislocation and increasing intensity of extreme hazard events, poverty and vulnerability to disaster are expected to increase.

Sources: Agrawala *et al.* (2003); Huq (2001); Huq and Khan (2006).

be taken forward, whilst being sensitive to the 'limits and uncertainties of the knowledge base'.

Patterns within nations

Whilst such assessments and international comparisons of exposure to adaptive capacity are crucial to laying out the global inequalities involved, climate inequalities can be scaled and assessed for other units of analysis, as noted earlier. Here intranational studies, looking at patterns of impact within nations, have connected to the spatial analysis tradition of environmental justice research. Such studies have argued that it is misleading to homogenise the population of whole countries as similarly able to cope with climate change, and that the types of differentiations in vulnerability and adaptive capacity predicted internationally can be mirrored *within* countries as well. Differentiations that have been discussed include by race, ethnicity and cultural group, income and gender (Terry 2009).

For example, a study of the social distribution of climate change impacts in the UK concluded that there could be considerable differentiation across the population as to who would be affected by changes in weather patterns, flooding, heat extremes, water shortage and sea level rise. The people most likely to be vulnerable to climate change are expected to be those that 'are already deprived by their health, the quality of their homes and mobility; as well as people who lack awareness of the risks of climate change, the capacity to adapt and who are less well supported by families, friends and agencies' (CAG Consultants 2009: iii). Income, age, gender and ethnicity are seen to interweave with these vulnerability characteristics.

A similar study undertaken in the US, but focused specifically on race, argues that African-Americans will be disproportionately affected by and vulnerable to climate change (Hoerner and Robinson 2008). Examples of evidence used to substantiate this claim include that:

- the six states with the highest African-American population are all in the Atlantic hurricane zone, and are expected to experience in the future more intense storms resembling Katrina and Rita;
- an increase in frequency and intensity of heatwaves or extreme heat events is expected and African-Americans suffer heat death at 150 to 200 per cent of the rate for non-Hispanic whites;
- in 2006, 20 per cent of African-Americans had no health insurance, which is nearly twice the rate among non-Hispanic whites, limiting their capacity to access treatment for health problems induced by climate change;
- the average income of African-American households is 57 per cent that of non-Hispanic whites, limiting access to resources for adaptation.

Other US studies provide some support and have generated further claims about likely inequalities in the impact of climate change. For example, Wilson *et al.* (2010) have attempted to map vulnerability to climate change impacts across the US at county level (whilst stressing the methodological complexities and uncertainties involved). By combining 39 variables, they generate vulnerability scores that reveal climate change vulnerability to be concentrated to some degree in the south of the country, with a larger number of people of colour, more exposure to the coasts and more poor people who are underserved medically. They also find high vulnerability in metropolitan areas in the Midwest and Northeast.

The extent to which the findings of such studies can be translated into other contexts, and always support the general claims of Hoerner and Robinson (2008) that 'global warming amplifies nearly all existing inequalities' (ibid.: 1), needs to be examined. The case of heatwaves may be instructive here. Epidemiological (and some anthropological) evidence collected from various parts of the world reveals only one common demographic characteristic in deaths from heatwaves – being old. When analysed by other socio-demographic characteristics, a mixed and inconsistent picture emerges – see Box 8.4. That is not to say that vulnerability to heatwaves is not differentiated along class, race or gender lines

---

Box 8.4 Who dies in heatwaves?

The most established and consistent pattern throughout the literature is that it is older people who regularly comprise the majority of victims in hot weather events, and this trend is demonstrated across a variety of locations and climates throughout the world. Within Europe older people have been found to be the most vulnerable group, for example, in studies focused on Madrid, Italy, France and the Netherlands among others. The scale of the increase is made clear by Fouillet *et al.*'s (2006) analysis of heat-related mortality during the 2003 heatwave in France. Of the 14,729 excess deaths that occurred in total, 11,731 were amongst people aged 75 and over, while only 2,930 were younger than 75. The same pattern has also been identified throughout the US, where north-eastern and northern interior cities with a cooler climate exhibit a much stronger pattern than do warmer cities in the south.

In studies examining differences by gender, the pattern of vulnerability is less clear. Both men and women have been shown to be at heightened risk in studies undertaken in different locations. In Europe a series of city and national studies have consistently shown that women comprise a higher percentage of the observed excess deaths than men. However, these findings are contradicted by studies conducted elsewhere. Amongst the US population either the pattern is reversed, with studies finding that men are more vulnerable than women (for example, Taylor and McGwin 2000), or no pattern is found at all.

In the US many studies have found that black people are more vulnerable to the adverse health effects of hot weather (for example, Medina-Ramon *et al.* 2006). Analysis has shown that it is unlikely that this is a result of racial differences in physiology; rather, it is a reflection of the lower socio-economic status and/or higher temperature exposure suffered by the black community (Basu and Samet 2002). Members of the black population are much less likely to have air-conditioning than the white population (O'Neill *et al.* 2005); lower socio-economic groups are more likely to live in higher heat stress inner-city neighbourhoods; and mortality has been shown to increase disproportionately in poor, urban, non-white areas. However, outside of the US, researchers have found that the impact of socio-economic status is less clear. Studies showed it to have no bearing on heat-related deaths in Australia and São Paulo, but a strong impact in Rome.

Source: Adapted and updated from Brown and Walker (2008).

amongst different older populations, but rather that the differentiation and vulnerability appear to be structured differently in different places with different physical and social conditions.

Having outlined various forms of evidence on inequalities in climate change impacts and how these are constituted, we can now turn to questions of justice through the lens of the three justice concepts introduced in Chapter 3 – distribution, procedure and recognition.

Adaptation and distributional justice

Most of the discussion so far has been inherently concerned with matters of distributional justice, as evidenced by unequal patterns in who will be affected and to what degree. In Chapter 3 a number of key questions addressed in distributive justice claims were identified, the first being the definition of who the recipients of justice are, or the community of justice. In the case of climate change, as already noted, the community of justice is being extended to encompass future generations, not just those alive in the here and now. Whilst climate change impacts, many would contend, are already being observed and adaptation is already being required, it is future generations that are expected to carry the major burden and therefore *intergenerational* distributional justice that is at issue.

Whilst the degree of current *and* future expected international or intranational inequalities in climate change impacts could, by themselves, be used to substantiate a claim of distributional injustice, for most observers climate injustice is structured through the disconnection between patterns of impacts and patterns of responsibility. As Roberts (2009: 185) put it:

> The core injustice of climate change – pointed out repeatedly by ethicists and representatives of the world's poorest nations in negotiating halls and on the floor of the U.N. General Assembly – is that those who are least responsible for causing the problem are also those most likely to suffer directly its early impacts. They should not, and simply cannot, pay for the massive expenses of preparing for, coping with, and recovering from increasing hurricanes, droughts, flooding, heatwaves, and sea level rise that climate change is bringing.

Climate change is in this way a clear example of where claims of injustice become particularly powerful where they combine together two patterns of distribution, one of harm or disbenefit and the other of responsibility for that harm. Other examples have been seen in preceding chapters on air quality and waste disposal. If miraculously it was the highest carbon emitters that were to suffer the severest consequences of climate change, then we could speculate that much might be different in the way that the problem and its solutions were viewed.

This 'double injustice' provides the basis of demands for resources and expertise to flow from those responsible for climate change to those at risk of its impacts, so that the latter are able to improve their capacities to adapt both now and in the future. Whilst without justice arguments financial flows from North to South might be viewed as a form of charity or assistance (Muller 2009), with justice firmly centre stage resource flows for adaptation become variously constituted as demands for the protection of human rights, as an enactment of the polluter pays principle, or as a form of compensation for damages inflicted. Each of these ways of making demands implies different forms of distributional outcome, indicating that resolving the distributional principle on which just adaptation policy should be constructed is not straightforward – both in terms of where the money should come from and how it should be distributed. Many different proposals embodying different justice principles have been made (Grasso 2009; Muller 2001), and, as with justice in general, we should expect the appropriate resolution to reflect the context involved, particularly when decisions are being made at a local level (Elster 1992).

For example, in the setting of adaptation decisions in the UK, such as investing in flood defences or protection against coastal erosion, Tompkins identifies five alternative principles for deriving adaptation justice (see Box 8.5). Each of these principles establishes different priorities in determining what is fair. The principle of 'reward', for example, prioritises a linkage between adaptation and mitigation, whilst 'equality' prioritises a simple notion of equal spatial allocation.

In the international UN agreements made to date on adaptation finance, some broad notion that justice should prioritise the most vulnerable (the 'deserving' or 'exposed' in Box 8.5) can be found, and there is a compelling case for this as an

---

Box 8.5 Alternative principles for justly distributing adaptation funding in the UK

Reward: We should reward those communities that are reducing their emissions the most.

Deserving: We should allocate the majority of the funds to those least able to adapt to climate change on their own.

Development: We should target 'failing' areas and use the adaptation funds as a lever for development.

Exposed: We should allocate to those who face direct impacts of climate change.

Equality: We should allocate adaptation funds equally to all areas in the UK.

Source: Tompkins, in Economic and Social Research Council (2010).

---

overarching principle. In the UN framework categorical distinctions are made between 'vulnerable' and 'particularly vulnerable' countries, and distinctions are also made between countries with particular physical characteristics, such as small island countries, countries with low-lying coastal areas, and countries with fragile ecosystems. Dow *et al.* (2006) make a strong case for giving the most vulnerable special attention, arguing that there is a moral imperative to do so. To justify their position they make a linkage to Rawls' (1972) second principle of a just society – one that is arranged so that the position of the least advantaged is optimised – but also to notions of justice as desert, capability, need and liberty, which they also see as each supporting their position. Prioritising the most vulnerable, they then argue, leads to a set of related ethical considerations as to how adaptation strategies should be carried out, including questions of timing, scope and process – see Table 8.1 for their full list.

Table 8.1 Ethical considerations in selecting adaptation strategies

Consideration	Preferred characteristics of adaptation strategy
Timing of implementation	Aim to avoid harm rather than to compensate after the fact
Scope of risk addressed	Both reduce potential impacts and address the causes of climate change
Burden transfer	Reduce risk rather than transfer it to other places, populations or the future
Liability	Include policies that create liability for risks that cannot be eliminated or removed
Scope of social injustice addressed	Avert harm and address social processes contributing to vulnerability
Capacity building	Increase the capacity of the most vulnerable to manage risks on their own
Human rights	Contribute to securing fundamental human rights and promoting social progress and better standards of life
Self-determination	Respect the right to self-determination through participatory processes that facilitate input into assessment of the relevance of vulnerability reduction adaptations as well as the range of options developed, offered and supported

Source: Derived from Dow *et al.* (2006).

Adaptation, recognition and procedural justice

The case for prioritising the most vulnerable readily connects justice as distribution to justice as procedure and recognition. As argued in Chapter 3, these three forms of justice, whilst conceptually distinct, are in practice readily intertwined. One of the most extreme forms of misrecognition is when a community or cultural group is under threat of its very survival. Whilst genocide is normally reserved as a term for deliberate acts of violence, particular place-based communities and cultural groups have argued that the impacts of climate change could amount to much the same thing and that their most fundamental of human rights are under threat. Two cases have been particularly prominent.

Small island communities, in some cases constituting whole 'small island states', will, because of sea level rise, see their homes, settlements, livelihoods and established ways of life effectively eradicated (Barnett and Campbell 2009; Kelman 2010). For these places and people, notions of adaptation become particularly acute and problematic as however much finance, expertise and knowledge is applied, only wholesale community, or even nation state, relocation provides any form of resolution. To do this, though, is arguably to fundamentally contravene their right to recognition. Detailed participatory research undertaken in Tuvalu (Mortreux and Barnett 2009; Paton and Fairbairn-Dunlop 2010), a state made up of nine low-lying islands, has shown how the depth of significant cultural, spiritual, familial and historical values that people have invested in the islands means that relocation would entail unbearable psychosocial losses. Many people as a consequence said they could not countenance leaving their island, others that the terms under which any migration (either as a family or as a community) takes place would be crucial.

A second case is that of communities based in the Arctic regions of northern Canada and Alaska. Here the warming trend predicted by climate models is at twice the rate of the global average with major implications for sea ice and permafrost cover, species distribution and associated hunting practices. Adger (2004) captures their need as 'the right to keep cold', one where the indigenous communities of the Arctic face fundamental threats to their ways of life and culture as a result of climate change impacts already being experienced. Using an environmental justice framing, Trainor *et al.* (2007) emphasise how the indigenous Northern Arctic communities' intimate relationship with their local ecosystem for physical and cultural sustenance is fundamentally put at risk by climate change. Whilst these communities are 'dynamic, learning entities' and have successfully adapted to various forms of threat and opportunity over the years, the scale and speed of climate-induced change is far beyond all previous experience. For example, the ringed seal is the staple food source and they depend upon stable shore-fast sea ice for breeding. Over the past 30 years, the average extent of sea ice has declined significantly; climate change scenarios project an acceleration of this melting and an almost ice-free Arctic Ocean within the next 100 years. With this key food source severely depleted, it is very hard to see how adaptation to an alternative means of sustenance can take place.

The particular vulnerabilities of and threats to such communities also raise questions of procedural justice – how they and others should be involved in decision-making about adaptation. The last entry in Table 8.1 above stresses 'self-determination' and the right to engage in participatory processes that assess and decide on adaptation options, and this has been a constant demand of those campaigning for the rights of the most vulnerable populations, from local through to global scales. For example, the charity Oxfam International (2008: 12), in examining how human rights need to be protected in climate change processes, argues for the need to:

> ensure that the most affected populations and social groups have effective voice in setting mitigation targets and policies. Countries with populations at greatest risk – such as the least-developed countries (LDCs), small-island developing states, and those in sub-Saharan Africa – must be allowed to participate fully and have effective voice in international negotiations on mitigation. Organisations representing indigenous people, women, and children, must also be able to participate effectively, nationally and internationally.

Globally, particular groups of nations, such as the small island states, have forcefully argued for their right to be heard in international negotiation processes, over and beyond the limited voice they would normally expect to have in such arenas. In practice, such processes are not though accessed or experienced with any sort of meaningful equality between all nations (Roberts and Parks 2007). Local adaptation processes also raise key issues of procedural justice. Who is to decide how and when local adaptation is to take place? What forms of adaptation are 'best' and how is the choice between the types of justice or fairness principles listed in Box 8.5 to be made? Again, there are recurrent demands for local, inclusive, community involvement, with Oxfam International (2008: 18) demanding that 'states must ensure that the most affected communities participate in, and have ownership of, the design and implementation of adaptation initiatives in order to safeguard their rights'. Adger and Barnett (2009: 2803) argue that community involvement is necessary because of the way that the social context in which adaptation takes place shapes values:

> Communities value things differently and these must be taken into account if adaptation is to be effective, efficient, legitimate, and equitable … Thus, what may be perceived as a successful adaptive response from a policy point of view may not be perceived the same way by those who have presumably benefitted from the activity.

Agyeman *et al.* (2009) recount the case of a north Alaskan Eskimo village which is already planning to move its 400-year-old settlement 18 miles further inland because of melting permafrost and rapid erosion of beaches and cliffs. Despite this severe disruption to a community whose livelihood is based almost entirely on the sea, the residents have not been opposed to the plan. They suggest that this is because 'all 650 villagers were involved in discussions with politicians and scientists about how to best manage the impending threat of the rising tide and were given the opportunity to democratically approve the managed retreat plan by 2/3 majority vote' (ibid.: 511). In Tavalua, where local participation has to date been much less substantial, Paton and Fairbairn-Dunlop (2010) emphasise the particular local context in which some forms of participation are seen as far more meaningful and relevant than others – traditional structures of the family, church and local government and local oral networks being rated by local people as far more relevant than more distant forums or technology-based mechanisms. Local participatory opportunities also rest on the wider national context. Trainor *et al.* (2007) show how in the Northern Arctic vulnerable indigenous communities in Canada are in a better institutional position to have an influence on adaptation strategies than those within the US, where under the Bush administration there was resistance to enabling anything more than the undertaking of scientific assessments of potential impact scenarios. Indigenous groups have been less visible and less recognised in natural resource decision-making.

Whether relying on procedural justice will always produce optimal adaptation outcomes is open to question, however, and there can be unintended consequences. For example, Adger and Barnett (2009) point to the possibility that talk of relocation can lead to local confidence being eroded and to rapid economic and social stagnation and blight. They also note the opportunities that the possibility of transformation presents to powerful actors, who may use climate change as a cover to achieve forced migrations for political or economic gain (Barnett and Campbell 2009). Few *et al.* (2007) contend that the uncertainties and long time scales involved in adaptation present particular challenges to normative principles of participation. In the context of coastal adaptation in the UK, they argue that the practice of participation has to learn from existing critiques of participatory processes in other contexts and that real tensions are likely to be felt between principles of democratic participation and the demands of 'anticipatory adaptation'. Rather than follow an 'illusion of inclusion', it may be better for participatory approaches to be more narrowly instrumental and for clarity about the scope and limitations of public involvement to be made explicit from the outset. As in other contexts, therefore, *exactly* what participation can deliver in both process and outcome terms, and from whose perspective, is a key issue. This observation carries into the realm of climate change mitigation to which we now turn.

Mitigation, responsibilities and transitions

Mitigation aims to reduce greenhouse gas emissions in order to limit the extent and speed of global warming and the scale of climate change impacts. Climate change cannot now entirely be prevented, but it can be limited, raising questions about how much mitigation should be aimed for and by when, and how it should be achieved and by whom. These difficult questions have been at the core of the contentious politics of climate change, primarily because taking action to curb emissions involves shifts away from established ways of doing things (for example, producing energy, moving around), including, as stressed earlier, ways of generating profit and wealth. Significant mitigation also implies the commitment of major amounts of private and public expenditure. It is not surprising, therefore, that there are such protracted global debates about who should shoulder the responsibility to act (distributional justice), and that these are interwoven with how decisions on setting mitigating targets and committing mitigation resources should be made (procedural justice).

Justice in mitigation is not though only about the assignment of responsibilities for mitigation. It is also about the distributional impacts of taking the mitigation actions themselves. Here we need to be concerned about both the direct and indirect justice implications of mitigation policy, its known consequences and its known and unknown or hidden 'side effects'. As we shall see, depending on how mitigation is pursued, the potential side effects of carbon and GHG emissions reductions may include lost jobs and reduced incomes, higher energy prices, food shortages and slower economic growth, with very uneven impacts on different parts of the world and different social groups. Alternatively, carbon mitigation actions may have far more productive 'side effects' which have the potential to promote forms of sustainable development, generate jobs and improve quality of life. Much therefore rests on exactly how carbon mitigation is pursued, and how transitions towards a low carbon society are conceived and implemented.

As in the case of adaptation, most attention has been given to the justice of mitigation at a global level, and it is at this scale that the discussion to follow begins. We will then move to the sub-national and local scale, which is where the unequal consequences of mitigation policy are most directly experienced.

Global targets and carbon mitigation

As with global adaptation policy, there are closely related procedural and distributional justice issues that arise within global mitigation policy. The procedural justice issues are in many ways the same as those discussed for adaptation – matters of representation, comparative power and voice and the resourcing of involvement in international negotiation processes. This section therefore focuses mainly on questions of distribution and responsibility as embodied within the setting of climate change targets.

We saw earlier how a key challenge of climate justice was the contestation around how the world is divided up into negotiating and target-setting blocks. The base unit used is the nation state and it is justice between nations (or blocks of nations) which is at issue for global climate justice. As also noted earlier, much rests in climate justice claim-making on how evidence is deployed, i.e. how greenhouse gas emissions are calculated, what indicators are used and what is and isn't taken account of within these. To provide some initial context, Figures 8.3 and 8.4 use two such indicators – average per capita CO_2 emissions and total CO_2 emissions – in order to convey something of the enormous absolute and relative variation in emissions that exists between nations in different parts of the world and the differences in country ranking between these two measures. China, for example, now has the highest total emissions but sits somewhere in the middle of the scale for emissions per person.

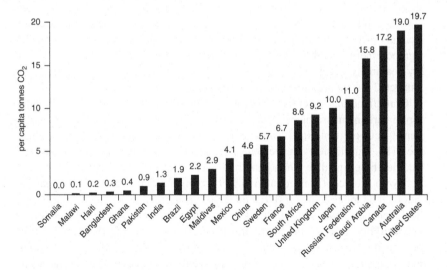

Figure 8.3 Per capita emissions of CO_2 in 2006 for selected countries.
Source: Data from UN Statistics Division, http://unstats.un.org/

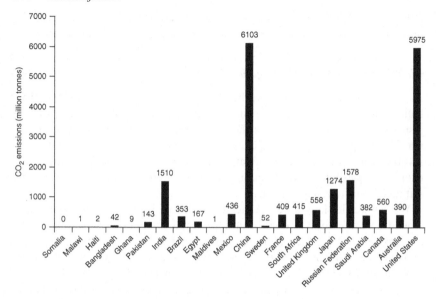

Figure 8.4 Total emissions of CO$_2$ in 2006 for selected countries.
   Source: Data from UN Statistics Division, http://unstats.un.org/

Across the parties to international negotiations and the literature on climate justice, both total emission and per capita data, as well as other evidence of GHG emissions, are brought into justice claim-making, promoting quite different understandings of how responsibility should be assigned and fairness enacted. Roberts and Parks (2007), who provide one of the most systematic and extensive accounts of these debates, identify four main ways of 'cutting the carbon cake' – grandfathering, carbon intensity, per capita and historical responsibility. The key features of each of these approaches, identifying forms of evidence and connected justice principles, are summarised in Table 8.2.

'Grandfathering' essentially accepts much of the current degree of differentiation in carbon emissions between countries, seeing justice in terms of an entitlement to current emissions levels and a proportionate reduction from this base line. The significant evidence here is total emissions at, and in relation to, a specified base-line date. Grandfathering was the approach effectively adopted in the Kyoto Protocol (as negotiated in 1997), which required the world's wealthier nations (and highest emitters) to reduce their emissions in percentage terms from the level they were at in 1990. The principle of 'common but differentiated responsibility' framed this international agreement, meaning that some nations took greater percentage cuts than others, whilst some were allowed to increase their emissions. The horse-trading that went on between negotiating teams allowed some countries to argue that their distinctive national economic context meant that their economic performance would suffer unduly if significant cuts

Table 8.2 Four different approaches to determining greenhouse gas emissions targets

	Approach	Key form of evidence	Justice principle
Grandfathering	Start from the emission levels that countries already have and seek to achieve proportional reductions from this base line	Total emissions at a base-line date	Entitlement to what is produced; proportional equality (nations are unequal and should be treated unequally); pragmatism
Carbon intensity	Focus targets on making economic growth less carbon intensive	Greenhouse gas emissions per unit of economic activity	Economic cost effectiveness, maximising overall utility in economic terms
Per capita	Focus on average emissions per person (calculated at a national level) and work towards making these more equal	Average per capita emissions at a current or future date	Egalitarian, equal rights to the atmosphere
Historical responsibility	Take account of the accumulated past emissions that have contributed to previous, current and future warming	Historic emissions levels (absolute and/or per capita) as well as current	As for per capita, plus compensation or reparation for past emissions

Source: Summarised from Roberts and Parks (2007).

were imposed equally on all. The US stayed entirely out of the convention, justified by President Bush on the grounds that 'the American way of life is not negotiable'. In this way, concern about the consequential side effects of carbon mitigation amongst powerful global actors has within the Kyoto Convention, and much subsequent debate, dominated over taking responsibility for current and future climate change impacts. This is a position that climate justice campaigners see as amoral and unjust, even more so if less developed countries with currently low emissions were also expected to follow a grandfathering principle (see later discussion).

'Carbon intensity' enrols a different form of evidence, measuring greenhouse gas emissions relative to economic activity and using this as the key indicator for setting targets. This approach was pushed strongly by the US government after it withdrew from the Kyoto Convention, arguing that it was a cost-effective way of achieving greenhouse gas reductions whilst also enabling economic growth – a narrow economically-framed utilitarian view of the justice of emission targets. The US set itself its own target using this measure of achieving an 18 per cent cut in GHG *intensity* over a ten-year period, expecting this to be achieved through the adoption of energy efficiency measures, clean technologies and carbon sequestration. Whilst advocates have argued that focusing on GHG intensity is efficient at a macro scale and good for developing nations, as it provides for their economic growth, critics have argued that it is a way of pretty much carrying on business as usual, with no cap set on total future emissions. As with grandfathering, it also embodies an assumption of entitlement to the stock of emissions that are already being produced.

The 'per capita' approach selects a third indicator, emissions per person averaged across a nation. This takes the total emissions of a country (including industry, transport, public sector emissions) and divides this figure by the total population of the country to produce an average. It is therefore not a measure of the emissions each person is directly responsible for (for example, from their own energy bills, car and air travel), but rather a collectively assigned equalised responsibility for the total emissions of all the greenhouse gas emitting activity within the nation. The per capita approach has been widely supported by those working within a climate justice framing and by various alignments of non-OECD nations, who argue that it embodies a simple egalitarian notion of justice that everyone has an equal right to the global atmosphere. Whilst approaches based on per capita evidence vary in their detail, the 'contraction and conver-gence' model has been most widely publicised and supported (Meyer 2000). This begins by setting a target for the stabilisation of the concentration of greenhouse gas emissions in the atmosphere at a certain future date (for example, 450 parts per million of carbon dioxide by 2040), one which is intended to avoid the worst effects of climate change. What that means in terms of the total emissions that can be allowed in order to achieve that stabilisation target is then calculated. This total is then divided by the world's population to derive a per capita target that all countries will move towards. As shown in Figure 8.5, the 'contraction' is therefore a reduction in global emissions, the 'convergence' a move towards equal per capita emissions by a specified date. What this means in practice is far more radical reductions in emissions by rich developed countries, with high per capita emissions, whilst poorer developing countries, with far lower per capita emissions, are able to increase their emissions to the global average.

In justice terms, the 'contraction and convergence' model both implements the notion of an equal right to the atmosphere and enables developing countries to take forward economic growth and investment in the energy, transport and other infrastructures that are needed to support development and improved quality of life. As with grandfathering, the side-effect consequences of carbon mitigation are important in how justice is understood here, but crucially the focus is on the consequences for poorer developing countries and their need for development, rather than on the economic interests of the globally rich and powerful. Whilst immediately compelling, there have been a range of criticisms of the per capita formulation of climate justice, including that it would be economically inefficient and politically impossible to achieve, having far too damaging impacts on the economies and societies of wealthy nations. In terms of justice theory, Starkey (2008) has also questioned whether a simple equal per capita allocation is

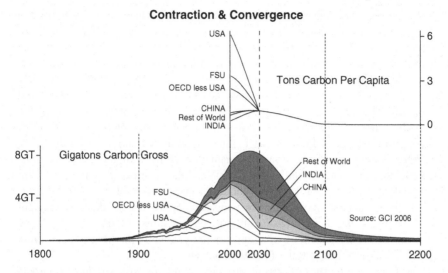

Figure 8.5 Contraction and convergence of greenhouse gas emissions.
Source: Global Commons Institute, http://www.gci.org.uk/index.html

supported by any contemporary philosophical traditions, challenging both the notion that the atmosphere is a 'commons' that is collectively owned, and to which all therefore have rights, and the implication that people also have an equal entitlement to the consumption of energy.

'Historical responsibility' makes a further critique of the per capita approach and indeed all the others. This is that, by only focusing on current and future emissions, the legacy that still remains from past historic emissions is neglected – sometimes also referred to as an ecological debt. Organisations such as Friends of the Earth International and negotiating blocks of poorer nations advocate an approach that *does* look back in time to factor in historic emissions on the grounds that this provides for fair reparation or compensation for past and ongoing impacts. This approach is also seen as a full enactment of the polluter pays principle, in that greenhouse gas reductions are set by taking full account of a country's contribution to global temperature increases (Roberts and Parks 2007). The need to stretch temporalities in this way is explained by the fact that carbon dioxide, in particular, stays in the atmosphere for a long time (100–120 years), with the consequence that most of the carbon emitted during the twentieth century is still in the atmosphere and contributing to changes in the climate. What this means in practice is that further pressure is put on those countries that industrialised first and have developed their economies over an extended period, and less pressure on (or more space for) developing countries that have a minimal historic contribution to be taken account of. Not surprisingly, this approach has been very contentious.

Reviewing these four approaches – which simplify what is a far more diverse field of proposed methods of target setting – we can see how different forms of evidence and different forms of justice concept are wrapped up together. The justice claims that are constructed (implicitly or explicitly) have to fit together evidence and justice concept in order to have some coherence, but they are clearly also deeply political configurations designed to promote particular national and international interests; in very crude terms the developed world interests of the G7/G20 countries are set against the developing world interests of the G77, although in fact the political scene is far more complex and nuanced than this, with the Copenhagen Summit in 2010 displaying shifting alliances and positions (see Box 8.1 for some flavour of this). Whilst climate justice advocates start from the political position that broadly egalitarian justice principles should be paramount (hence per capita or historic responsibility approaches), other framings put justice in the background, focusing far more on questions of efficiency, economic growth and the realpolitik of what appears pragmatically possible or achievable given the current distribution of political power.

The deep complexities of setting and then achieving carbon reduction targets also reflect the historical and contextual embedding of patterns of pollution and development, explored in Chapter 5 in relation to local air pollution. The same embedding is true of greenhouse gas emissions and is reflected in the notion of 'common but differentiated responsibility' in the UN convention – the recognition that some nations are in a more difficult position when it comes to making emissions cuts than others and that 'differentiation' is therefore needed when setting targets. In this respect, the side-effect consequences of carbon mitigation need to be taken seriously, for both developed and developing countries, not least because of their potential to further disadvantage the already disadvantaged in quite significant ways – see later discussion. This all suggests, as Roberts and Parks (2007: 150) put it, that some form of hybrid solution to deriving a fair profile of emission reductions will be required:

> what is needed is moral compromise, or a negotiated justice settlement. Strict adherence to particularistic notions of justice, by comparison, is a perfect recipe for a stalemate. Therefore it will likely be an optimal mix of principles that will enable rich and poor nations to overcome barriers to cooperation.

Rescaling: within-nation justice in mitigation responsibilities

Much is clearly at stake in the global negotiation of climate change targets. However, what really matters in the end are the changes that are *actually* achieved within nations to make emission reductions, from national and regional government down to the local level of communities, homes and business. Justice matters here as well. Looking *within* aggregate totals, per capita averages and broad-brush generalisations about the status of particular countries reveals further sets of inequalities and further questions about the distribution of responsibilities for, and side-effect consequences of, greenhouse gas mitigation. In this sense climate justice is not only relevant internationally. We can also 'rescale' climate justice concerns to examine their intranational profile (Davies and Kirwan 2010).

Focusing first on responsibilities, various studies have begun to produce finer-grained evidence of the contribution that different social groups are making to carbon emissions. Most have focused on variation by income or wealth. For example, in a study of the patterns of distribution of household carbon dioxide (CO_2) emissions across income groups in four European countries (Netherlands, UK, Sweden and Norway), Kerkof *et al.* (2009) found that emissions increased by a factor of three to four with increasing income in all four countries (despite the profiles of emissions levels and intensities being quite variable). In other words, across all four countries those with the highest incomes are responsible for three to four times the emissions of the poorest. Brand and Boardman (2008) focus specifically on personal travel-related emissions in the UK. Their study is based on a detailed survey of travel behaviours in one part of the UK, which is then translated into emissions data and aggregated up; despite the methodological limitations, their analysis provides a broad indication of the scale of the inequalities involved. As shown in Figure 8.6, the highest income group is responsible

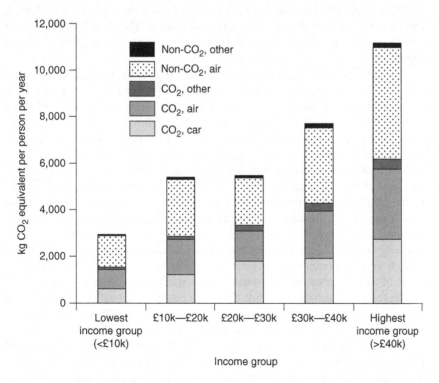

Figure 8.6 Transport mode breakdown of average greenhouse gas emissions per person by income group, disaggregated by CO_2 and non-CO_2 emissions.
Source: Adapted from Brand and Boardman (2008).

for 3.5 times the annual travel emissions of the lowest income group, with the top 10 per cent of emitters responsible for 43 per cent of total GHG emissions, whilst the bottom 10 per cent are responsible for only 1 per cent. They also found significant relationships between total emissions and being in work, gender (with older women having particularly low emissions), age (with 36–65 year olds disproportionately responsible for emissions), household composition and size (single households having the highest per capita emissions largely because of air travel) and car ownership. This suggests a rather complex social profiling of carbon emissions, although one in which income is a strongly differentiating factor.

Such evidence suggests significant problems with assigning a collective and equal responsibility for climate change to all citizens within a country such as the UK. Whilst the level of inequality in emissions is nothing like as vast as it is at a global level, and emissions of the lowest income groups in the UK are still higher than those of the poorest in other parts of the world, it is still possible to argue for the notion of 'common but differentiated responsibility' to be applied at an intra-national as well as international scale. Following this line of reasoning, Brand and Boardman (2008) argue that their study of travel emissions means that certain sub-groups of the population who are disproportionately responsible for emissions should be actively targeted in policy – the 'taming of the few' as they put it. It is also important to note that even in the poorest countries in the world, there can be elite groups with very high carbon emissions. There are both 'Souths in Norths and Norths in Souths', to borrow from a more general point about the profile of inequalities across the world.

This leads us to the question of the strategies, interventions, methods and tools by which climate change mitigation, or more broadly low carbon transition, is to be pursued (Davies and Kirwan 2010). There are a great variety of possibilities and alternatives, including multiple technological innovations and deployments, changes in patterns of supply (of energy, water, food etc.) and interventions that are intended to achieve changes in patterns of consumption and everyday life. As well as each method having different (and often disputed) degrees of effectiveness and efficiency, each targets responsibility to a greater or lesser degree on particular actors, with implications in terms of fairness in relation to their responsibility for GHG emissions. To some degree we can see a mirroring of the different approaches taken at a global level. For example, initiatives such as the 10:10 campaign in the UK, which seeks to get individuals, businesses and organisations to sign up to a common 10 per cent reduction in emissions, are essentially a form of 'grandfathering'. Personal carbon trading, in contrast, allocates all people (or more probably all adults) within a country (or smaller spatial unit) an equal allowance of carbon, which they can then trade with others – buying from others if they need a bigger carbon allowance (for example, to make multiple international flights) or selling to others if they have an allocation that is unused (Paterson and Stripple 2010). Although there are deep problems with how such a system would work in reality, and some have voiced ethical objections (Caney 2010), it is in principle an attempt to embody a 'per capita' climate justice principle in a national or sub-national carbon mitigation mechanism.

Double dividends and regressive 'side effects'

As well as the justice embodied in the assignment of responsibilities, we also again have to be very aware of the sub-national scale side effects of carbon mitigation policy and how these are socially distributed. There are major issues here about how interventions of different forms will lead to price increases, job losses, opportunities perhaps for some and negative consequences in terms of wellbeing, quality of life and health for others. There has been much discussion about how certain climate mitigation strategies and interventions might provide the opportunity to produce progressive social and economic outcomes, ones that bring benefits to people who are already disadvantaged. For example, Meadowcroft (2008: 333) argues that climate change can provide a socially progressive window of opportunity:

> By responding to problems related to climate change, social policy may be able to advance other objectives related to welfare and equity. Just as some analysts talk of the policy response to climate change as representing an 'economic opportunity agenda' (rather than just an economic cost agenda), so it can also be considered a 'social opportunity agenda'.

At an international level, arrangements such as offset markets and the Clean Development Mechanism (CDM) are intended to do just that, focusing low carbon or carbon sequestration investments in developing countries, with the intention of bringing local economic and social benefits to disadvantaged communities as well as global benefits in terms of carbon mitigation. Whether or not the reality of such projects on the ground has always been to produce such a 'double dividend' is contested, and there have been strong critiques of particular local experiences. Lohmann (1999), for example, argues that the working of carbon offset markets means that forestry projects which claim to soak up carbon dioxide will be located where land is cheapest and that therefore it will be the poorest who will have to make room for them. Efficient production of these new carbon sinks also means that forest plantations will tend to be monocultures 'either in the sense of species monocultures or in the sense of environments which are socially simplified from above to serve a single purpose'. Worldwide experience with such plantations, he argues, has shown damaging effects for livelihoods and democratic politics. Boyd (2009) examines two examples of CDM forest projects in South America, contrasting the narratives of elite organisations with local narratives of those on the ground. Elite actors, she argues, embody a simplified portrayal of complex natural ecosystems and of the human dimensions of global environmental change, with tree offset projects conducted largely from the top down. Local narratives, she found, were far more ambiguous and reflected conflict, uncertainty and opposition, and a distribution of benefits which did not readily materialise for those most in need.

Within developed nations, much campaigning attention has also been given to the possibilities of a synergy between climate mitigation and social justice or

Table 8.3 Examples of potential distributional justice issues arising from carbon mitigation measures

Mitigation measure	Description	Potential distributional justice issues
Carbon tax	A tax on electricity which reflects the carbon content of the method of generation, incentivising shifts to low carbon sources and greater efficiency	Increased energy prices potentially exacerbates fuel poverty problems for vulnerable groups – those that are poor and struggling to pay bills, older people, and those in low-quality energy inefficient housing.
Biofuels	The growing of biofuels as a lower carbon form of fuel, usually for transport	Biofuels replace crops grown for food, leading to food shortages and price rises. Large corporate actors take over land for biofuels previously used for small-scale farming and local markets.
Cap and trade	Large-scale carbon emitters are given a maximum emission allocation, but can buy more emissions rights by trading with other emitters	Existing polluters are able to buy credits to sustain emissions of both carbon and other more locally damaging pollutants. Populations living near to polluting industries tend to be poor and in some contexts predominantly from racial/ethnic minorities. Benefits go to large companies.
Nuclear new build	New nuclear power stations are built as a low carbon energy source, replacing fossil fuel generation	Nuclear power stations concentrate the risks of accidents/leaks on local populations, and on other communities across the nuclear fuel cycle. They introduce risks for future generations that have to deal with the legacy of decommissioning and management of radioactive wastes for thousands of years.
Carbon capture and storage	The capturing of carbon emissions from power stations and their storage underground to prevent accumulation in the atmosphere	The scale of any risks of harmful releases from carbon distribution and storage systems are uncertain but are locally concentrated. These risks will extend between generations as long-term storage is likely to be required.

poverty policies, in the form of a new 'Green Deal' that can generate green employment, for example. In Europe the trade unions have been active in pushing such an agenda, arguing for new alliances between green and social groups and for a renewing of the EU sustainable development strategy to take up the opportunity for a more socially and environmentally just formulation (Degryse and Pochet 2009). Whilst these are impressive attempts to be proactively optimistic about the doing of justice within climate policy, there are real tensions about how much it might be possible to achieve. And there are real risks that climate policy might in fact worsen inequalities and impose further burdens on those that are already disadvantaged; see Preston and White (2010) for an analysis of the distributional impacts of UK climate policies. Table 8.3 takes five examples of carbon mitigation measures and outlines the potential distributional justice implications arising from their implementation. In reading this table, we might well remember the global-scale climate justice gains that are involved for vulnerable nations and communities, and how these are then to be 'balanced' in some way against the negative distributional consequences realised at other scales.

Towards integration in climate justice

With so many justice dimensions involved – adaptation, mitigation, global, local, distributive, procedural and more besides – some form of integration between them is needed. Following the generic arguments made earlier in this book, distributive, procedural and recognition notions of justice cannot be disentangled and so need to be pursued simultaneously. As Paavola *et al.* (2006: 264) succinctly put it, 'redistribution without empowerment can be shortlived and empowerment without redistribution can be an insult'. At another level, the overall balance between mitigation and adaptation is itself a justice issue. To take strong mitigation action now is to limit the need for adaptation in the future. To be weaker on mitigation is to increase the need for future adaptation. At each end of this mitigation–adaptation spectrum, and across the many positions in between, the profile of uneven (and uncertain) distributional implications and outcomes (who pays, who suffers, who decides, etc.) shifts radically in spatial, temporal and social terms. It would be unwise to suggest that we can now, at this point in time, derive some perfectly determined and integrated formulation of climate justice, in part because of the challenges of knowledge, scale and power outlined at the beginning of this chapter. Clearly, in the context of climate change seeing climate justice as a dynamic process of 'working towards' (see Chapter 9) makes particular sense.

Following this rationale, what we can draw from the academic and campaigning literature are lists of key principles or priorities that cover both mitigation and adaptation. Paavola *et al.* (2006), for example, suggest four cornerstones of fair climate change policy:

- *avoiding dangerous climate change* by setting a safe maximum level for greenhouse gas concentrations in the atmosphere that reflects the low capacity of vulnerable systems to adapt;
- *forward-looking responsibility* through implementing a carbon tax or similar scheme which generates funds to compensate for the impacts of the climate changes that are not avoided;
- *putting the most vulnerable first* and understanding vulnerability in terms of not just exposure but also the capacity to cope and adapt;

Table 8.4 Rights-centred principles for climate policy

Human rights principles for policy-making	Mitigation – reducing greenhouse gas emissions: essential to respect and protect human rights	Adaptation – building resilience to unavoidable impacts: now essential as a remedy for failing to respect and protect human rights
Guarantee a core minimum – a basic standard of rights for all	States must implement national and international mitigation targets and policies that minimise the risk of exceeding 2°C warming	States must target disaster relief and adaptation initiatives to safeguard the essential claims – to life, food, water, shelter and health – of the most vulnerable people
Focus on vulnerability and those whose rights are most at risk	States must ensure their mitigation policies do not undermine vulnerable people's rights, domestically or overseas	States must ensure that support for adaptation is channelled to the most vulnerable communities, such as women, minority groups and children
Ensure participation of people whose rights are affected by policies	States must ensure that the most affected communities and groups have effective voice in setting national and international mitigation targets and policies	States must ensure that the most affected communities participate in, and have ownership of, the design and implementation of adaptation initiatives in order to safeguard their rights
Provide accountability and remedies for violations	States must report publicly on results in implementing mitigation targets and policies	States must ensure effective and transparent governance of national and international adaptation strategies and funds
Deliver on international co-operation to realise rights worldwide	States must take on emissions cuts in line with their national responsibility for causing climate change and their capability to assist	States must finance international adaptation based on their national responsibility for causing climate change and their capability to assist

Source: Oxfam International (2008).

- *fair participation for all* which encompasses participation at multiple levels and across multiple forms of actor.

Oxfam International (2008) expresses a more differentiated list – see Table 8.4 – in the form of a set of human rights principles which are then applied to both mitigation and adaptation equally to form a symmetrical and comprehensive scheme. A tough set of demands on nation states is laid out here.

As a further example we can look to a formulation of policy principles rooted more explicitly in an environmental justice frame. The '10 Principles of Just Climate Change Policies' (see Box 8.6) drawn up by the Environmental Justice and Climate Change (EJCC) initiative are positioned from a US perspective and draw together both global and far more local justice dimensions, focusing on global responsibilities as well as the notion of a 'just transition' for workers and communities and monitoring the distributional consequences of carbon markets.

Much of course rests on the degree to which such formulations of just climate principles can have an influence in practice and lead through to shaping actions

---

Box 8.6 The EJCC '10 Principles of Just Climate Change Policies in the U.S.'

Stop Cooking the Planet. Global climate change will accelerate unless we can slow the release of greenhouse gases into the atmosphere. To protect vulnerable Americans, we must find alternatives for those human activities that cause global climate change.

Protect and Empower Vulnerable Individuals and Communities. Low-income workers, people of color, and Indigenous Peoples will suffer the most from climate change's impact. We need to provide opportunities to adapt and thrive in a changing world.

Ensure Just Transition for Workers and Communities. No group should have to shoulder alone the burdens caused by the transition from a fossil fuel-based economy to a renewable energy-based economy. A just transition would create opportunities for displaced workers and communities to participate in the new economic order through compensation for job loss, loss of tax base, and other negative effects.

Require Community Participation. At all levels and in all realms, people must have a say in the decisions that affect their lives. Decision-makers must include communities in the policy process. U.S. federal and state governments, recognising their government-to-government relationship, must work with tribes as well.

Global Problems Need Global Solutions. The causes and effects of climate change occur around the world. Individuals, communities, and nations must work together cooperatively to stop global climate change.

The U.S. Must Lead. Countries that contribute the most to global warming should take the lead in solving the problem. The U.S. is four percent of the world's population but emits 25 percent of the world's greenhouse gases. All people should have equal rights to the atmosphere.

Stop Exploration for Fossil Fuels. Presently known fossil fuel reserves will last far into the future. Fossil fuel exploration destroys unique cultures and valuable ecosystems. Exploration should be halted as it is no longer worth the cost. We should instead invest in renewable energy sources.

Monitor Domestic and International Carbon Markets. We must ensure that carbon emissions and sinks markets are transparent and accountable, do not concentrate pollution in vulnerable communities, and avoid activities that harm the environment.

Caution in the Face of Uncertainty. No amount of action later can make up for lack of action today. Just as we buy insurance to protect against uncertain danger, we must take precautionary measures to minimize harm to the global climate before it occurs.

Protect Future Generations. The greatest impacts of climate change will come in the future. We should take into account the impacts on future generations in deciding policy today. Our children should have the opportunity for success.

Source: http://www.ejcc.org/.

in real terms, particularly in the US where climate policy has only recently begun to move in more progressive directions. In this respect, Sze *et al.* (2009) provide an interesting commentary on the recent development of climate change policy in California, where environmental justice groups managed to exert some influence over the formulation of a key legislative bill, resisting the deployment of mandated 'cap and trade' market mechanisms and requiring various participatory processes to be followed. Long-standing tensions between local environmental justice activists and mainstream national environmental groups emerged, in part because of the different scales at which impacts and outcomes were understood. Conflict over the implementation of the bill continues. As just one example, albeit from a place that produces greenhouse gas emissions at a faster rate than almost anywhere else in the world, this emphasises both the hard-fought achievements that can be made, as well as the many obstacles that stand in the way of making just climate policy a reality.

Summary

We have seen in this chapter how the framing of climate change as a global risk and universal problem, which we are all part of and collectively responsible for, has to be strongly tempered by a recognition of the deep patterns of difference and inequality involved. These permeate our responsibility for and attempts to mitigate climate change through emission reduction, and our attempts to cope with and adapt to its impacts. In both respects, evidence of inequality and justice claims exist at global/international scales as well as intranational/local scales, and balancing these, or, better, addressing them simultaneously, is far from straightforward. Doing justice in the here and now can also conflict with doing justice to future generations – making justice part of the choices to be made between action on mitigation and action on adaptation. Putting these various dimensions of justice together, what is fundamentally at issue is the form of sustainable societal transition we want to pursue as we seek to move towards a low carbon and climate adapted future. A high carbon future is likely to produce intolerable burdens that are focused on the most vulnerable around the world. A low carbon future *could* be a more environmentally and socially just one – however these terms are conceived – and advocating climate justice necessarily has to work towards such an objective.

Further reading

For much authoritative material on the science and impacts of climate change and on policy issues see the websites of the Intergovernmental Panel on Climate Change (IPCC, www.ipcc.ch) and of the Tyndall Centre (www.tyndall.ac.uk). Key campaigning websites on climate justice include:

- www.climate-justice-action.org
- www.climatejustice.org.uk
- www.actforclimatejustice.org
- www.climatejusticefund.org

The 'contraction and convergence' approach has a dedicated web resource at www.gci.org.uk/index.html. There have been various recent UK and European initiatives to explore the more local justice implications of climate change and energy transition, including those of the King Baudoin Foundation (www.kbs-frb.be), the Joseph Rowntree Foundation (www.jrf.org.uk/), the New Economics Foundation (www.neweconomics.org/) and the InCluESEV project (http://incluesev.kcl.ac.uk/). Cornerhouse provides reports taking a radical and critical perspective on various aspects of climate change policy (www.thecornerhouse.org.uk/).

Key books covering the justice and fairness implications of adaptation and mitigation include Adger *et al.* (2006), Roberts and Parks (2007), Vanderheiden (2008) and Arnold (2010). *Climate Ethics: Essential Readings*, edited by Gardiner *et al.* (2010), provides an excellent collection of important material. A special issue of the journal *Environmental Justice* (2009, vol. 2, no. 4) focuses on climate justice and of *Gender and Development* (2009, vol. 17, no. 1) on climate change and gender.

9 Analysing environmental justice
Some conclusions

I started this book with an open and wide-ranging perspective on environmental justice as being concerned with the intertwining of environment and social difference. Having now moved in preceding chapters through much material – ideas, concepts, empirical evidence and arguments – it is evidently the case that environment and social difference are intertwined in uneven ways, such that some benefit while others do not, and that some have power and access to influence over environmental outcomes, whilst others are excluded. In other words, like much else in the contemporary world, the environment is socially differentiated and unevenly available. To some degree this is evident through everyday observation. Take a city like New York, that I am currently visiting; it is very clear that the lives of some of its residents involve an extraordinary level of access to and consumption of resources, including environmental ones, and a mobility which means that the continual noise, pollution and concrete of downtown can be routinely replaced with the quiet, clean air and nature of 'out of town' and up-state. For other New Yorkers the quality of their lives, health and well-being is so very different, for many reasons, but including because of the environmental burdens they live with, and the limited essential resources they can obtain. Follow the media reporting of disasters such as the devastating floods, landslides and earthquakes which, over the period of writing this book, have happened in Pakistan, Haiti, Australia, Japan, New Zealand, Brazil and many other places, and it is patently evident that these events are not experienced in even ways, that some places and populations are better protected, some more vulnerable and others more resilient.

However, I have argued in this book that it is necessary to go beyond these sorts of observations, however immediate and compelling they might be, to examine in a more detailed, analytical and critical way the discourse of environmental justice, to separate out and examine the different elements of environmental justice claim-making, to think carefully about evidence of inequalities, how it is constituted and what it means. By critics of the academic practice of making everything more complicated than it appears to need to be, and of using concepts and theories when common sense might appear sufficient, this 'taking apart' of environmental justice may be given short shrift. However, I would argue that by looking back over preceding chapters we can find many reasons why developing a thoughtful and critically informed view of environmental justice is worthwhile. These reasons relate not only to academic advancement but also, more importantly, to knowing better the interrelation between environment and social difference and therefore to more effectively pursuing the improvement of the conditions under which people are able to lead their lives and the way they are treated in environmental terms. To complete this discussion, therefore, I have identified seven conclusions, and also assertions, that emerge from and consolidate the learning I hope has been achieved through the book and which advocate ways of taking forward the analysis of environmental justice in the future.

There is value in understanding exactly how social differentiation exists and how it is experienced in environmental terms

We have seen that whilst at times broad generalisations about who is burdened by risk and pollution and who is not, who lives with waste, who has access to greenspace and so on, can be sustained by empirical analysis, the breadth of evidence we have reviewed suggests that socially differentiated patterns can be more complex and varied than such generalisations allow for. To make this point is not to seek to undermine the general claims made by the environmental justice movement, but rather to point out the need to target our attention on where patterns of inequality are most problematic and where they do most matter. The fact that some studies in some cities or regions have shown that greenspace is equally shared across social groups, or even that there is better provision for poorer communities than richer ones, is something to be welcomed, and something to be learnt from in seeking to replicate elsewhere. Similarly, we saw in Chapter 5 that evidence suggests that sometimes, in some places, air pollution does not follow the poor, that the worst air quality within a region or city can be found in areas that are the least deprived. This again does not undermine other studies in other places that have found diametrically opposed results. It rather leads us to think carefully about why this counterintuitive social differentiation appears to exist, and how we might need to evaluate it in a way that brings to bear considerations such as available residential choice and access to mobility.

We also need to recognise that social differentiations intersect, so that in Chapter 6 it was clear that Hurricane Katrina did not just have the most severe impacts on the poor black population of New Orleans, but that it was the older members of this population that were most vulnerable and most likely to die, sometimes in the most appalling circumstances. We also saw how the nature of flood impacts, and who they most affected, shifted to some degree across the disaster cycle, so that tracing the dynamics of social differentiation over time was important. In the case of climate change, the need to recognise the multiple forms that climate impacts can take, how these can be identified at different scales of analysis, and the uncertainties involved in any predictive analysis of social differentiation, also demands a careful multi-dimensional and multi-scaled analysis. In these and other cases, therefore, being specific, looking across multiple forms of social difference and targeting our attention accordingly have to be important. Otherwise there is a real danger that particular vulnerabilities and particular exclusions will remain invisible, lost in the broad sweep of generalisation about how environmental justice plays out.

Environmental inequalities are constituted by more than spatial patterns of proximity and exposure

I have argued at length in a previous paper (Walker 2009a) that the dominant preoccupation of much empirical environmental justice research with spatial patterns of proximity (to hazardous facilities, to greenspace), or with residence in more or less polluted spaces, works with a limited and restricted geography, and also with an insufficient sense of what can really matter to the health and well-being of populations. Spatially structured statistical studies are an important starting point, but they are often insufficient on their own. Following this line of argument, we have seen in several of the preceding chapters that analysis of social differentiation needs to go further to focus also on who is more or less vulnerable to the impacts of environmental bads, on who is more or less in need of environmental goods, and on matters that are less susceptible to conventional modes of empirical analysis. Chapter 5 on air quality stressed the 'triple jeopardy' of deprived populations being potentially most exposed to air pollution, most vulnerable to suffering ill health as a consequence, and least able to cope with that ill health, and the need to develop a more holistic view of their life-worlds. Chapters 6 and 8 on flooding and climate change both stressed the need to understand patterns of vulnerability and ability to cope, recover and adapt. Chapter 7 on greenspace argued that some populations may be more in need of the health and well-being benefits of greenspace than others, with the consequence that uneven patterns of greenspace that are biased towards these populations might be exactly the goal that is to be sought after, rather than some form of spatial equality.

These examples show the value in being aware of the multiple patterns involved, that these can and do overlap and compound each other (for example, unequal exposure compounded by unequal vulnerability), and therefore the importance of building this into environmental justice analysis. Similar points were also made about patterns of responsibility for environmental bads and their impacts – such as responsibility for waste production, for air pollution, for carbon emissions – and how these need to be set against, and often in contrast to, patterns of exposure and vulnerability in arriving at judgements about the nature and depth of injustice that exists.

Recognising the methodological complexities and choices involved in generating empirical evidence is important in progressing understanding and in doing justice

The two points made so far make considerable demands on the practices of evidence or knowledge generation, and throughout the preceding chapters I have sought to make clear the methodological complexities, choices, scale dependencies, assumptions and uncertainties that are involved in undertaking varieties of empirical environmental justice research. My intention in positioning evidence as a form of claim-making was to promote careful, critical thinking (not the same as criticism in its always negative, colloquial form) about the detail of how environmental justice evidence and arguments are put together, and to raise awareness of the strengths and weaknesses of *any* body of evidence that is drawn on. This means that some evidence may be more highly valued and given more significance than other evidence, that improvements in the evidence base can and should be made over time, and that alternative methodologies can be experimented with that range right across the qualitative–quantitative spectrum. Increasingly we are seeing such moves in environmental justice scholarship, which is being greatly enriched as a consequence.

In discussing matters of evidence, the processes through which this is generated have also been at issue. Indeed, one of the constituent elements of procedural justice listed in Chapter 3 was inclusion in and access to research. This is one of the many ways in which evidence and justice are interrelated (hence the integrated framework of Chapter 3), meaning that evidence can be seen to embody or carry justice and not just inform the content of our justice judgements. This means that giving attention to the ethics of how research is carried out, and who is involved in setting priorities, in making methodological choices and decisions and in interpreting and making sense of results, needs to be part of evidence-generation. This does not mean that all environmental justice research should be community-based, participatory and entirely open at its outset, as this would limit the scope and scale of what can be examined and achieved and rub up against many practical and institutional constraints. But such moves, which do present a significant challenge to much scientific practice, should be made where it makes sense and where they can be.

It is necessary to distinguish between inequality and injustice and to reason carefully about why an inequality matters and to whom

The distinction between inequality and injustice, description and normative evaluation, was introduced in the first chapter and has run through subsequent chapters, providing a way of holding back from the immediate and simple judgement that anything that is unequal is also wrong. In each of the substantive chapters there has been a working through of various grounds on which justice arguments may be developed, or the resolution of a just condition might be formulated. Sometimes this has involved integrating in matters of vulnerability, need and responsibility, as already argued, and sometimes also matters of choice, agency and process. There certainly has not been one formulation of what constitutes environmental justice that provides a meaningful resolution across the topics of waste, air quality, flooding, greenspace and climate change. Within each of these domains various alternative resolutions have been identified.

There have also been some difficult tensions to deal with, such as between making a judgement of 'what is just' based on a distributional outcome – for example, the location of a waste dump or reprocessing operation – and making one on the basis of the process through which that outcome was derived. This classic distinction in justice reasoning plays out in many environmental justice contexts, and also raises questions about who is to judge what is to be valued and what is just. For example, we saw that for greenspace an external judgement that experiencing nature and using greenspace are 'good things' needed to be tempered by recognition of the other ways that nature and greenspace might be evaluated, and that there are therefore dangers of imposing particular cultural values in making justice judgements. There are depths of resources to draw on here in normative and political theory, but the applied working through of such considerations is also important given the close linkage there can be between the way that evidence is collected and utilised (for example, top down or bottom up) and the basis on which value and justice judgements are made. These points also play into the understanding of environmental justice framing developed in Chapter 2, in the way that the key concerns of environmental justice activism, research and policy are not entirely constant but are situated in the place and time in which they are manifest. The US EPA's view and definition of environmental justice, for example, may well work for them but be insufficient and inappropriate for anti-nuclear focused activists working in Taiwan. There are commonalities, yes, but there are also important differences in the sense that is made of the two words this book has been concerned with.

Environmental justice is about more than just patterns of distribution; procedure, recognition and their detail, also matter

The point that environmental justice in practice is multivalent and draws from different traditions of justice theory emerges out of the arguments already made. However, in utilising the work of Schlosberg (2004, 2007), and then applying these ideas across the chapter topics, hopefully it has become clearer why taking on board more than distribution, and recognising its integration with matters of procedure and recognition, make so much sense. This is undoubtedly at the core of the arguments and claims made by environmental justice activists, as Schlosberg has argued, but it is also productive in analysing particular concerns, with climate change being a paramount example. To talk of climate justice as only a distributional issue would be to neglect its integrative procedural justice components, related to how mitigation and adaptation are conceived and practised, and the need for meaningful recognition of the most climate-vulnerable groups. The forms of tabulation that have been used at various points through the chapters – such as Table 4.1 comparing waste case studies and Table 6.2 analysing Katrina – have provided a simple tool for identifying justice claims in distributional, procedural and recognition terms, which could be deployed in taking forward the analysis of other cases. The related point here is that there is a need for research that better deals with the procedural and recognition dimensions of environmental justice, using research methods that are more likely to be qualitative, experiential and participatory rather than involving the crunching together of statistical data sets. Such research is needed, because the detail, in particular of procedural justice, does matter. As we have seen in several situations, whether or not a process of participatory decision-making leading to an environmental outcome with social consequences can stand up to scrutiny as suitably inclusive, legitimate, respected and deliberative (if those are the key criteria) rests very much on the detail of how it is carried out, rather than on the simple intent to have such a process in place.

Environmental justice is contested and involves political challenges

It is sometimes possible for researchers to get sucked into thinking that taking forward environmental justice is essentially about producing better evidence, formulating clearer arguments and developing better policy prescriptions. However important these might be, environmental justice activists clearly know that in practice making change happen is about politics, often fighting tooth and nail and challenging the way that power is exercised. This book has not centred very directly on the practical political work of environmental justice activism, but I hope it has recognised throughout that there is politics *always* involved – in the way that evidence is produced and utilised in public debate, in the way that climate change negotiations are carried out, in the way that the management of greenspace is passed over from state to voluntary bodies, in the way that Katrina victims were left to fend for themselves, in the way that the environmental justice movement in the US has been attacked by corporate-funded neoliberal think tanks, and so on. In a few places I have also reminded the reader that we cannot simply presume some common egalitarian frame of understanding about what matters, what is wrong and what should be done about it. There are counter-frames based on some very different normative orientations. For example, in Chapter 4 we examined the hard-line economist argument that patterns in who lives near to a waste site are just an outcome of choice and market processes, and in Chapter 6 the libertarian view that post-Katrina New Orleans demonstrates that less state involvement is needed in flood and disaster management rather than more. I have not lingered on such perspectives, but they clearly exist and are often influential in shaping the way that government and private-sector organisations at different levels set priorities and make decisions. And indeed they provide at least some of the explanation as to why environmental inequalities and environmental injustice have been produced in the first place.

The inclusion of the explanatory processes through which environmental inequalities are produced within the framework laid out in Chapter 3 stressed the need to understand the 'how and why' of environmental injustice rather than its only surface manifestation. This discussion distinguished between 'contextual process claims' about particular explanations in a place and time and 'structural process claims' which provide an overarching theoretical perspective on how environmental inequalities are produced. Whilst such explanations are crucial if injustice is to be addressed, they can build up to daunting if not overwhelming challenges; for example, when the deep historical embedding of contemporary socio-spatial patterns are revealed, or when the many dimensions of power exercised by the 'polluter-industrial complex' (Faber 2008) or through the accumulation processes of the treadmill of production (Gould *et al.* 2004) are laid bare. Here though I have been impressed by the way in which many of the environmental justice activists that I have talked to are very able and ready to describe the fundamental structures and processes they are up against (often in clearer and more eloquent ways than practising academics), but are still able to see the often small, local and incremental advances they can achieve as important and worthwhile. The political challenges of advancing environmental justice and shifting the terms of debate are substantial, perhaps most substantially and significantly so in the case of climate change, but this should encourage rather than dissuade the development of better reasoning, better argument and more compelling articulations of what is at issue.

Environmental justice is an objective but also a process of 'working towards'

This sense of the ongoing work involved in pursuing environmental justice brings us to a final point and argument about how environmental justice is defined and understood. In the opening chapter of the book various selected definitions of environmental justice were given, drawn from academic, activist and policy literatures. Each of these specified environmental justice as an objective, something that is to be sought after, articulating various visions of equality, fairness, health and well-being. It was argued that using an objective-based definition is an important part of constructing a politically powerful environmental justice frame, and that an objective also does the crucial job of providing a metric against which current conditions can be judged and critiqued.

However, imagine, for a moment, a world in which 'everyone, everywhere' lived with equally good environmental quality, equally good access to environmental resources, equal influence over environmental decision-making, and equal respect as a member of society. A nirvana maybe, but not one that is properly definable or achievable in anything except the abstract. Dig below the surface simplicity of these imagined conditions and, as we have seen throughout the preceding chapters, we find enormous complexities involved in defining the practical meaning of 'equality' and how justice in these terms might be operationalised. Not only that, but to conceive of an environmentally just world is to enter into a fiction in which all kinds of inequalities in political, economic and cultural power are also neatly and finally resolved. In this light, defining environmental justice in terms of objectives and desired conditions might be productive in various practical and political ways, but it doesn't inherently recognise the active work, contestation and dynamics involved in working *towards* these objectives in a constantly evolving and shifting environmental, economic, social and political context. Similar arguments have been made about the inherent dynamics and ambivalences also to be found in pursuing sustainability and the need to continually debate and renegotiate what this means (Becker and Jahn 1999; Walker and Shove 2007).

We can therefore think about environmental justice as a dynamic process as well as an objective, a process of working towards environmental justice objectives and against forces which are serving to produce or sustain patterns of injustice, in short as 'work in progress'.[1] Conceiving environmental justice in this way is not standard in the existing academic literature, although a number of people have made moves in this direction. Pellow (2000), for example, has made the case for a process-based understanding of the ways in which environmental inequalities and injustices are created, capturing both the ongoing dynamics and the contestations between actors that are involved. But we can also think about action towards revealing, challenging and countering injustice as dynamic and ongoing, a process which includes the use of environmental justice as a frame for collective action and political activity, and the making of claims about patterns of inequality and the justice or otherwise of these. Iris Marion Young characterises this productive process of claim-making as follows:

> appeals to justice and claims of injustice are not a result, they do not reflect an agreement; they are rather the starting point of a certain kind of debate. To invoke the language of justice and injustice is to make a claim, a claim that we together have obligations of certain sorts to one another. Many listening to the claim will disagree about precisely what those obligations are or how they should be met, but as long as we are arguing about what is just in this situation we are acknowledging we are together politically and owe at least minimal commitments of solidarity to one another.
>
> (Young 1998: 40)

A process perspective fits with an understanding of justice which recognises that the resolution of the questions 'what is just?' and 'what is good?' will never, and should never, be finally resolved, but will be continually open to reasoning, revision and challenge. A more open and dynamic understanding of environmental justice does not imply that there cannot be agreements, progress and resolutions of problematic situations along the way. But these will never finally resolve inequality and injustice always and forever, and in any case the terms in which these situations are understood will be dynamic rather than static and frozen in time.

---

1 I have drawn this specific phrase from a talk given by Cecil Corbin Mark of West Harlem Environmental Action to a group of students. I am grateful for his articulation and the understanding that lies behind it.

Bibliography

Adger, N., 2004. The right to keep cold. *Environment and Planning A*, 36(10): 1711–15.

Adger, N. and Barnett, J., 2009. Four reasons for concern about adaptation to climate change. *Environment and Planning A*, 41: 2800–5.

Adger, N., Paavola, J., Huq, H. and Mace, M.J. (eds), 2006. *Fairness in Adaptation to Climate Change*. Cambridge, MA: MIT Press.

Agarwal, S. and Brunt, P., 2006. Social exclusion and English seaside resorts. *Tourism Management* 27(4): 654–70.

Agrawala, S., Ota, T., Ahmed, A.U., Smith, J. and van Aalst, M., 2003. *Development and Climate Change in Bangladesh: Focus on Coastal Flooding and the Sundarbans*. Paris: OECD.

Agyeman, J., 1987. Black people in a white landscape: social and environmental justice. *Built Environment*, 16(3): 232–6.

Agyeman, J. and Evans, T., 2003. Toward just sustainability in urban communities: building equity rights with sustainable solutions. *Annals of the American Academy of Political and Social Science*, 590: 35–53.

Agyeman, J. and Evans, B., 2004. 'Just sustainability': the emerging discourse of environmental justice in Britain? *Geographical Journal*, 170(2): 155–64.

Agyeman, J., Bullard, R. and Evans, B. (eds), 2003. *Just Sustainabilities: Development in an Unequal World*. London: Earthscan.

Agyeman, J., Devine-Wright, P. and Prange, J., 2009. Close to the edge, down by the river? Joining up managed retreat and place attachment in a climate changed world. *Environment and Planning A*, 41: 509–13.

Akaba, A., 2004. Science as a double-edged sword: research has often rewarded polluters, but EJ activists are taking it back. *Race, Poverty, Environment*, XI(2): 9–11.

Alexandri, E. and Jones, P., 2008. Temperature decreases in an urban canyon due to green walls and green roofs in diverse climates. *Building and Environment*, 43(4): 480–93.

Alwang, J., Siegel, P.B. and Jorgensen, S.L., 2001. *Vulnerability: A View from Different Disciplines*. Washington, DC: World Bank.

Anderton, D., Anderson, A., Oakes, J.M. and Fraser, M., 1994. Environmental equity: the demographics of dumping. *Demography*, 31(2): 229–48.

Angus, I., 2009. *The Fight for Climate Justice*. San Francisco, CA: Resistance Books.

Arnell, N.W., 2002. *Hydrology and Global Environmental Change*. London: Pearson Education.

Arnold, D., 2010. *The Ethics of Global Climate Change*. Cambridge: Cambridge University Press.

Association of British Insurers, 2002. *The Flood and Coastal Defence Funding Review. The Association of British Insurers: Response to the Defra and National Assembly for Wales Consultation Document.* London: ABI.

Baden, B.M. and Coursey, D.L., 2002. The locality of waste sites within the city of Chicago: a demographic, social, and economic analysis. *Resource and Energy Economics*, 24(1–2): 53–93.

Bankoff, G., Frerks, G. and Hilhorst, D., 2004. *Mapping Vulnerability: Disasters, Development and People.* London: Earthscan.

Bard, D., Laurent, O., Filleul, L., Havard, S., Deguen, S., Segala, C., Pedrono, G., Riviere, E., Schillinger, C., Rouil, L., Arveiler, D. and Eilstein, D., 2007. Exploring the joint effect of atmospheric pollution and socioeconomic status on selected health outcomes: an overview of the PAISARC project. *Environmental Research Letters*, 2(4): 045003, doi: 10.1088/1748–9326/2/4/045003.

Barnett, C. and Scott, D., 2007. Spaces of opposition: activism and deliberation in post-apartheid environmental politics. *Environment and Planning A*, 39: 2612–31.

Barnett, J. and Campbell, J., 2009. *Climate Change and Small Island States: Power, Knowledge and the South Pacific.* London: Earthscan.

Basel Action Network, 2002. *Exporting Harm: The High-Tech Trashing of Asia.* Seattle, WA: Basel Action Network.

Basel Action Network, 2005. *The Digital Dump: Exporting Re-use and Abuse to Africa.* Seattle, WA: Basel Action Network.

Basu, R. and Samet, J.R., 2002. Relation between Elevated Ambient Temperature and Mortality: A Review of the Epidemiologic Evidence. *Epidemiologic Reviews*, 24(2): 190.

Bates, L.K. and Green, R.A. (eds), 2009. *Housing Recovery in the Ninth Ward.* Boulder, CO: Westview.

Beck, U., 1998. *World Risk Society.* Cambridge: Polity Press.

Becker, E. and Jahn, T. (eds), 1999. *Sustainability and the Social Sciences.* London: Zed Books.

Beckerman, W. and Pasek, J., 2001. *Justice, Posterity and the Environment.* Oxford: Oxford University Press.

Beckman, L., 2008. Do global climate change and the interest of future generations have implications for democracy? *Environmental Politics*, 17: 610–24.

Beckman, L. and Page, E., 2008. Perspectives on justice, democracy and global climate change. *Environmental Politics*, 17: 527–35.

Been, V., 1994. Locally undesirable land uses in minority neighbourhoods: disproportionate siting or market dynamics. *Yale Law Journal*, 106(6): 1383–1422.

Been, V. and Gupta, F., 1997. Coming to the nuisance or going to the barrios? A longitudinal analysis of environmental justice claims. *Ecology Law Quarterly*, XXIV(1): 1–49.

Bell, D., 2004. Environmental justice and Rawls' difference principle. *Environmental Ethics*, 26(3): 287–306.

Bell, S., Hamilton, V., Montarzino, A., Rothnie, H., Travlou, P. and Alves, S., 2008. *Greenspace and Quality of Life: A Critical Literature Review.* Stirling: Greenspace Scotland.

Benford, R., 2005. The half-life of the environmental justice frame: innovation, diffusion and stagnation, in Pellow, D.N. and Brulle, R.J. (eds), *Power, Justice and the Environment: A Critical Appraisal of the Environmental Justice Movement.* Cambridge, MA: MIT Press, 37–54.

Benford, R. and Snow, D.A., 2000. Framing processes and social movements: an overview and assessment. *Annual Review of Sociology*, 26: 11–39.

Bickerstaff, K. and Agyeman, J., 2010. Assembling justice spaces: the scalar politics of environmental justice in north-east England, in Holifield, R., Porter, M. and Walker, G.P. (eds), *Spaces of Environmental Justice.* Chichester: Wiley, 193–218.

Bickerstaff, K. and Walker, G.P., 2005. Shared visions, unholy alliances: power, governance and deliberative processes in local transport planning. *Urban Studies*, 42(5): 2123–44.

Birkmann, J. (eds), 2006. *Measuring Vulnerability to Natural Hazards*. New York: United Nations University Press.

Blaikie, P. and Brookfield, H., 1987. *The Political Economy of Soil Erosion*. London: Methuen.

Block, W., 2006. Katrina: private enterprise, the dead hand of the past and weather socialism: an analysis of economic geography. *Ethics, Place and Environment*, 9(2): 231–41.

Block, W. and Whitehead, R., 1999. The unintended consequences of environmental justice. *Forensic Science International*, 100: 57–67.

Blowers, A., 1999. Nuclear waste and landscapes of risk. *Landscape Research*, 24(3): 241–64.

Boardman, B., Bullock, S. and McLaren, D., 1999. *Equity and the Environment*. London: Catalyst Trust.

Boholm, A. and Lofstedt, R. (eds), 2005. *Facility Siting: Risk, Power and Identity in Land Use Planning*. London: Earthscan.

Bolin, B., Nelson, A., Hackett, E.J., Pijawka, K.D., Smith, C.S., Sicotte, D., Sadalla, E.K., Matranga, E. and O'Donnell, M., 2002. The ecology of technological risk in a Sunbelt city. *Environment and Planning A*, 34(2): 317–39.

Bolund, P. and Hunhammar, S., 1999. Ecosystem services in urban areas. *Ecological Economics*, 29(2): 293–301.

Bond, P., 2000. Economic growth, ecological modernization or environmental justice? Conflicting discourses in South Africa today. *Capitalism, Nature, Socialism*, 11(1): 31–61.

Bowen, W., 2002. An analytical review of environmental justice research: what do we really know? *Environmental Management*, 29(1): 3–15.

Bowen, W. and Wells, M.V., 2002. The politics and reality of environmental justice: a history and considerations for public administrators and policy makers. *Public Administration Review*, 62(6): 688–98.

Bowen, W., Salling, M.J., Haynes, K.E. and Cyran, E.J., 1995. Toward environmental justice: spatial equity in Ohio and Cleveland. *Annals of the Association of American Geographers*, 85(4): 641–63.

Boyd, E., 2009. Governing the clean development mechanism: global rhetoric versus local realities in carbon sequestration projects. *Environment and Planning A*, 41: 2380–95.

Brainard, J.S., Jones, A.P., Bateman, I.J., Lovett, A.A. and Fallon, P.J., 2002. Modelling environmental equity: access to air quality in Birmingham, UK. *Environment and Planning A*, 34: 695–716.

Brand, C. and Boardman, B., 2008. Taming of the few: the unequal distribution of greenhouse gas emissions from personal travel in the UK. *Energy Policy*, 36: 224–38.

Briggs, D., Abellan, J.J. and Fecht, D., 2008. Environmental inequity in England: small area associations between socio-economic status and environmental pollution. *Social Science and Medicine*, 67(10): 1612–29.

Brighouse, H., 2004. *Justice*. Cambridge: Polity.

Brown, J.D. and Damery, S.L., 2002. Managing flood risk in the UK: towards an integration of social and technical perspectives. *Transactions of the Institute of British Geographers*, 27: 412–26.

Brown, P., 1995. Race, class, and environmental health – a review and systematization of the literature. *Environmental Research*, 69(1): 15–30.

Brown, P., Mayer, B., Zavestoski, S., Luebke, T., Mandelbaum, J. and McCormich, S., 2003. The health politics of asthma: environmental justice and collective illness experience in the US. *Social Science and Medicine*, 57: 453–64.

Brown, S. and Walker, G., 2008. Understanding heat wave vulnerability in nursing and residential homes. *Building Research and Information*, 36(4): 363–72.

Brownlow, A., 2006a. An archaeology of fear and environmental change in Philadelphia. *Geoforum*, 37: 227–45.

Brownlow, A., 2006b. Inherited fragmentations and narratives of environmental control in entrepreneurial Philadelphia, in Heynen, N., Kaika, M. and Swyngedouw, E. (eds), *In the Nature of Cities: Urban Political Ecology and the Politics of Urban Metabolism*. New York: Routledge, 208–25.

Brulle, R.J. and Pellow, D.N., 2005. The future of environmental justice movements, in Brulle, R.J. and Pellow, D.N. (eds), *Power, Justice and the Environment: A Critical Appraisal of the Environmental Justice Movement*. Boston, MA: MIT Press, 293–300.

Bryant, B., 1995a. *Environmental Justice: Issues, Policies and Solutions*. Covello, CA: Island Press.

Bryant, B., 1995b. Pollution prevention and participatory research as a methodology for environmental justice. *Virginia Environmental Law Journal*, 14: 589–611.

Bryant, B. and Hockman, E., 2005. A brief comparison of the civil rights movement and the environmental justice movement, in Pellow, D.N. and Brulle, R.J. (eds), *Power, Justice and the Environment: A Critical Appraisal of the Environmental Justice Movement*. Cambridge: MIT Press, 23–36.

Bryner, G.C., 2002. Assessing claims of environmental justice, in Mutz, K., Bryner, G.C. and Kenney, D.S. (eds), *Justice and Natural Resources: Concepts, Strategies and Applications*. Washington, DC: Island Press, 31–56.

Buckingham, S. and Kulcur, R., 2009. Gendered geographies of environmental injustice. *Antipode*, 41(4): 659–83.

Buckingham, S., Reeves, D. and Batchelor, A., 2005. Wasting women: the environmental justice of including women in municipal waste management. *Local Environment*, 10(4): 427–44.

Buckle, P., 1998. Re-defining community and vulnerability in the context of emergency management. *Australian Journal of Emergency Management*, 13(4): 21–6.

Buckle, P., Marsh, G. and Smale, S., 2000. New approaches to assessing vulnerability and resilience. *Australian Journal of Emergency Management*, 15(2): 8–14.

Bulkeley, H. and Walker, G., 2005. Environmental justice: a new agenda for the UK. *Local Environment*, 10(4): 329–32.

Bullard, R., 1983. Solid waste and the black Houston community. *Sociological Inquiry*, 53: 273–89.

Bullard, R., 1990. *Dumping in Dixie: Race, Class and Environmental Quality*. Boulder, CO: Westview Press.

Bullard, R., 1992. Environmental blackmail in minority communities, in Bryant, B. and Mohai, P. (eds), *Race and the Incidence of Environmental Hazards: A Time for Discourse*. Boulder, CO: Westview Press, 82–95.

Bullard, R., 1999. Dismantling environmental racism in the USA. *Local Environment*, 4(1): 5–19.

Bullard, R. (ed), 2005. *The Quest for Environmental Justice: Human Rights and the Politics of Pollution*. San Francisco, CA: Sierra Club Books.

Bullard, R. and Wright, B. (eds), 2009. *Race, Place, and Environmental Justice after Hurricane Katrina: Struggles to Reclaim, Rebuild, and Revitalize New Orleans and the Gulf Coast*. Boulder, CO: Westview.

Bullard, R., Mohai, P., Saha, R. and Wright, B., 2007. *Toxic Wastes and Race at Twenty: 1987–2007 Grassroots Struggles to Dismantle Environmental Racism in the United States*. Cleveland, OH: United Church of Christ.

Bullard, R., Johnson, G.S. and Torres, A.O., 2009. Transportation matters: stranded on the side of the road before and after disasters strike, in Bullard, R. and Wright, B. (eds), *Race, Place and Environmental Justice after Hurricane Katrina: Struggles to Reclaim, Rebuild, and Revitalize New Orleans and the Gulf Coast*. Boulder, CO: Westview, 63–88.

Burby, R.J., 2006. Hurricane Katrina and the paradoxes of government disaster policy: bringing about wise governmental decisions for hazardous areas. *Annals of the American Academy of Political and Social Science*, 604: 171–91.

Burgess, J., 1988. 'But is it worth taking the risk?' How women negotiate access to urban woodlands: a case study, in Ainley, R. (ed.), *New Frontiers of Space, Body, and Gender*. New York: Routledge, 114–28.

Burgess, J., Harrison, C.M. and Limb, M., 1988. People, parks and the urban green: a study of popular meanings and values for open spaces in the city. *Urban Studies*, 25: 455–73.

Burns, P.F. and Thomas, M.O., 2008. A new New Orleans? Understanding the role of history and the state–local relationship in the recovery process. *Journal of Urban Affairs*, 30(3): 259–71.

Buzzelli, M., 2007. Bourdieu does environmental justice? Probing the linkages between population health and air pollution epidemiology. *Health and Place*, 13(1): 3–13.

Byrne, J. and Wolch, J., 2009. Nature, race, and parks: past research and future directions for geographic research. *Progress in Human Geography*, 33(6): 743–65.

Cable, S., Mix, T. and Hastings, D., 2005. Mission impossible? Environmental justice activists' collaborations with professional environmentalists and with academics, in Pellow, D.N. and Brulle, R.J. (eds), *Power, Justice and the Environment: A Critical Appraisal of the Environmental Justice Movement*. Cambridge, MA: MIT Press, 55–76.

CAG Consultants, 2009. *Differential Social Impacts of Climate Change in the UK: Literature Review*, Project UKCC22. Edinburgh: SNIFFER.

Caney, S., 2007. Justice, borders and the cosmopolitan ideal: a reply to two critics. *Journal of Global Ethicism*, 3(2): 269–76.

Caney, S., 2010. Markets, morality and climate change: what, if anything, is wrong with emissions trading? *New Political Economy*, 15(2): 197–224.

Capek, S.M., 1993. The environmental justice frame – a conceptual discussion and an application. *Social Problems*, 40(1): 5–24.

Carmin, J. and Agyeman, J. (eds), 2011. *Environmental Inequalities Beyond Borders: Local Perspectives on Global Inequities*. Boston, MA: MIT Press.

Carruthers, D.V., 2008. The globalization of environmental justice: lessons from the U.S.–Mexico border. *Society and Natural Resources*, 21(7): 556–68.

Cerrell Associates Inc., 1984. *Political Difficulties Facing Waste-to-Energy Conversion Plant Siting*. Report prepared for the California Waste Management Board. Los Angeles, CA: Cerrell Associates Inc.

Chaix, B., Gustafsson, S., Jerrett, M., Kristerson, H., Lithman, T., Boalt, A. and Merlo, J., 2006. Children's exposure to nitrogen dioxide in Sweden: investigating environmental justice in an egalitarian country. *Journal of Epidemiology and Community Health*, 60: 224–34.

Chalmers, H. and Colvin, J., 2005. Addressing environmental inequalities in UK policy: an action research perspective. *Local Environment*, 10(4): 333–60.

Charles, A. and Thomas, H., 2007. Deafness and disability: forgotten components of environmental justice: illustrated by the case of Local Agenda 21 in South Wales. *Local Environment*, 12(3): 209–21.

Chauvin, S.W., DiCarlo, R.P., Lopez, F.A., Delcarpio, J.B. and Hilton, C.W., 2008. In for the long haul: sustaining and rebuilding educational operations after Hurricane Katrina. *Family and Community Health*, 31: 54–70.

Chavis, B.F.J., 1994. 'Preface', in Bullard, R. (ed.), *Unequal Protection: Environmental Justice and Communities of Color*. San Francisco, CA: Sierra Club Books, xi–xii.

Cigler, B.A., 2007. The 'big questions' of Katrina and the 2005 great flood of New Orleans. *Public Administration Review*, 67: 64–76.

Clapp, J., 2001. *Toxic Exports: The Transfer of Hazardous Wastes from Rich to Poor Countries*. Ithaca, NY: Cornell University Press.

Clapp, J., 2002. What the pollution havens debate overlooks. *Global Environmental Politics*, 2(2): 11–19.

Collins, T.W., 2010. Marginalization, facilitation, and the production of unequal risk: the 2006 Paso del Norte floods. *Antipode*, 42(2): 258–88.

Colls, J. and Tiwary, A., 2008. *Air Pollution*, 3rd edn. London: Routledge.

Comedia and Demos, 1995. *Park Life: Urban Parks and Social Renewal*. Stroud: Comedia.

Commission for Architecture and the Built Environment, 2009. *The Green Information Gap: Mapping the Nation's Green Spaces*. London: CABE.

Committee on Radioactive Waste Management, 2007. *Implementing a Partnership Approach to Radioactive Waste Management: Report to Governments, April 2007*, CoRWM document 2146, http://www.corwm.org.uk/ (accessed 2/4/08).

Commonwealth of Massachusetts, 2002. *Environmental Justice Policy*. Boston, MA: State House.

Corburn, J., 2002. Environmental justice, local knowledge, and risk: the discourse of a community-based cumulative exposure assessment. *Environmental Management*, 29(4): 451–66.

Coughlin, S.S., 1996. Environmental justice: the role of epidemiology in protecting unempowered communities from environmental hazards. *Science of the Total Environment*, 184: 67–76.

Curtice, J., Ellaway, A., Robertson, C., Morris, G., Allardice, G. and Robertson, R., 2005. *Public Attitudes and Environmental Justice in Scotland: A Report for the Scottish Executive on Research to Inform the Development and Evaluation of Environmental Justice Policy*. Edinburgh: Scottish Executive.

Cutter, S.L., 2006. The geography of social vulnerability: race, class, and catastrophe. *Understanding Katrina: Perspectives from the Social Sciences*, http://understanding katrina.ssrc.org/Cutter/ (accessed 20/9/10).

Cutter, S.L. and Solecki, W.D., 1996. Setting environmental justice in space and place: acute and chronic airborne toxic releases in the southeastern United States. *Urban Geography*, 17(5): 380–99.

Cutter, S.L., Mitchell, J.T. and Scott, M.S., 2000. Revealing the vulnerability of people and places: a case study of Georgetown County, South Carolina. *Annals of the Association of American Geographers*, 90(4): 713–37.

Cutter, S.L., Boruff, B.J. and Shirley, W.L., 2003. Social vulnerability to environmental hazards. *Social Science Quarterly*, 84(2): 242–61.

Damery, S., Petts, J., Walker, G. and Smith, G., 2008a. *Addressing Environmental Inequalities: Waste Management*. R&D Technical Report SC020061/SR3. Bristol: Environment Agency.

Damery, S., Walker, G., Petts, J. and Smith, G., 2008b. *Addressing Environmental Inequalities: River Water Quality*. Bristol: Environment Agency.

Daniels, R.J., Kettl, D.F. and Kunreuther, H. (eds), 2006. *On Risk and Disaster Lessons from Hurricane Katrina*. Philadelphia, PA: University of Pennsylvania Press.

Dasgupta, S., Huq, M. and Khaliquzzaman, M., 2006. Indoor air quality for poor families: new evidence from Bangladesh. *Indoor Air*, 16: 426–44.

Daunton, M.J. (ed.), 2000. *The Cambridge Urban History of Britain: 1840–1950*. Cambridge: Cambridge University Press.

Davies, A., 2006. Environmental justice as subtext or omission: examining discourses of anti-incineration campaigning in Ireland. *Geoforum*, 37(5): 708–24.

Davies, A. and Kirwan, N., 2010. *Rescaling Climate Justice: Sub-national Issues and Innovations for Low Carbon Futures*. IIIS Discussion Paper No. 340. Dublin: Institute for International Integration Studies.

Davis, B.C. and Bali, V.A., 2008. Examining the role of race, NIMBY and local politics in FEMA trailer park placement. *Social Science Quarterly*, 89(5): 1175–94.

Davis, M., 1998. *The Ecology of Fear: Los Angeles and the Imagination of Disaster*. New York: Metropolitan.

Dawson, J.I., 2000. The two faces of environmental justice: lessons from the eco-nationalist phenomenon. *Environmental Politics*, 9(2): 22–60.

Day, R., 2010. Environmental justice and older age: consideration of a qualitative neighbourhood-based study. *Environment and Planning A*, 42: 2658–73.

Day, R. and Wager, F., 2010. Park, streets and 'just empty space': the local environmental experiences of children and young people in a Scottish study. *Local Environment*, 15(6): 509–23.

Debbane, A.M. and Keil, R., 2004. Multiple disconnections: environmental justice and urban water in Canada and South Africa. *Space and Polity*, 8(2): 209–25.

Degryse, C. and Pochet, P., 2009. *Paradigm Shift: Social Justice as a Prerequisite for Sustainable Development*. Working Paper 2009.02. Brussels: European Trade Union Institute.

Delemos, J.L., 2006. Community-based participatory research: changing scientific practice from research on communities to research with and for communities. *Local Environment*, 11(3): 329–38.

Department of Health, 1998. *Quantification of the Health Effects of Air Pollution in the United Kingdom*. Department of Health Committee on the Medical Effects of Air Pollution. London: The Stationery Office.

Dichiro, G., 1992. Defining environmental justice: women's voices and grass-roots politics. *Socialist Review*, 22(4): 93–130.

Dixon, J. and Ramutsindela, M., 2006. Urban resettlement and environmental justice in Cape Town. *Cities*, 23(2): 129–39.

Dobson, A., 1998. *Justice and the Environment: Conceptions of Environmental Sustainability and Dimensions of Social Justice*. Oxford: Oxford University Press.

Dodds, L. and Hopwood, B., 2006. BAN waste, environmental justice and citizen participation in policy setting. *Local Environment*, 11(3): 269–86.

Dolan, A.H. and Walker, I.J., 2004. Understanding vulnerability of coastal communities to climate change related risks. *Journal of Coastal Research*, Special Issue 39.

Dow, K., Kasperson, R.E. and Bohn, M., 2006. Exploring the social justice implications of adaptation and vulnerability, in Adger, N., Paavola, J., Huq, H. and Mace, M.J. (eds), *Fairness in Adaptation to Climate Change*. Cambridge, MA: MIT Press, 79–98.

Duma, B., 2007. *Environmental Justice Networking Forum: An Organisation's Struggle to Regain Its Stronghold*, http://www.sangonet.org.za/portal/index.php?option=com_co ntent&task=view&id=4739&Itemid=374 (accessed 27/10/08).

Dunion, K., 2003. *Troublemakers: The Struggle for Environmental Justice in Scotland*. Edinburgh: Edinburgh University Press.

Dunion, K. and Scandrett, E., 2003. The campaign for environmental justice in Scotland as a response to poverty in a northern nation, in Agyeman, J., Bullard, R. and Evans, B. (eds), *Just Sustainabilities: Development in an Unequal World*. London: Earthscan, 311–22.

Dutcher, G.A., Spann, M. and Gaines, C., 2007. Addressing health disparities and environmental justice: the National Library of Medicine's Environmental Health Information Outreach Program. *Journal of the Medical Library Association*, 95(3): 330–6.

Economic and Social Research Council, 2010. *How Will Climate Change Affect People in the UK and How Can We Best Develop an Equitable Response?* Swindon: Economic and Social Research Council.

Eden, S., 1999. 'We have the facts': how business claims legitimacy in the environmental debate. *Environment and Planning A*, 31: 1295–1309.

Ederington, J., Levinson, A. and Minier, J., 2005. Footloose and pollution-free. *Review of Economics and Statistics*, 87(1): 92–9.

Elliott, J.R. and Pais, J., 2006. Race, class, and Hurricane Katrina: social differences in human responses to disaster. *Social Science Research*, 35(2): 295–321.

Elsom, D., 1996. *Smog Alert: Managing Urban Air Quality*. London: Earthscan.

Elster, J., 1992. *Local Justice: How Institutions Allocate Scarce Goods and Necessary Burdens*. New York: Russell Sage Foundation.

Enarson, E., 2006. Women and girls last? Averting the second post-Katrina disaster. *Understanding Katrina: Perspectives from the Social Sciences*, http://understanding katrina.ssrc.org/Enarson/ (accessed 20/9/10).

Endres, D., 2009. From wasteland to waste site: the role of discourse in nuclear power's environmental injustices. *Local Environment*, 14(10): 917–37.

English Nature, 1996. *A Space for Nature*. Peterborough: English Nature.

Environment Agency, 2004. *Addressing Environmental Inequalities: Position Statement*. Bristol: Environment Agency.

Environmental Justice and Climate Change Initiative, 2008. About EJCC, http://www. ejcc.org/ (accessed 28/11/09).

Evans, E., Ashley, R., Hall, J., Penning-Rowsell, E., Sayers, P., Thorne, C. and Watkinson, A., 2004. *Foresight: Future Flooding. Scientific Summary, Volume II: Managing Future Risks*. London: Office of Science and Technology.

Faber, D., 2005. Building a transnational environmental justice movement: obstacles and opportunities in the age of globalization, in Bandy, J. and Smith, J. (eds), *Coalitions across Borders: Transnational Protest and the Neoliberal Order*. Lanham, MD: Rowman and Littlefield, 43–68.

Faber, D., 2008. *Capitalizing on Environmental Injustice: The Polluter-Industrial Complex in the Age of Globalization*. Lanham, MD: Rowman and Littlefield.

Fairburn, J., Walker, G. and Smith, G., 2005. *Investigating Environmental Justice in Scotland: Links between Measures of Environmental Quality and Social Deprivation*, Report UE4(03)01. Edinburgh: Scottish and Northern Ireland Forum for Environmental Research.

Fairburn, J., Butler, B. and Smith, G., 2009. Environmental justice in South Yorkshire: locating social deprivation and poor environments using multiple indicators. *Local Environment*, 14(2): 139–54.

Fan, M.-F., 2006a. Environmental justice and nuclear waste conflicts in Taiwan. *Environmental Politics*, 15(3): 417–34.

Fan, M.-F., 2006b. Nuclear waste facilities on tribal land: the Yami's struggles for environmental justice. *Local Environment*, 11(4): 433–44.

Few, R., 2003. Flooding, vulnerability and coping strategies: local responses to a global threat. *Progress in Development Studies*, 3(1): 43–58.

Few, R., Brown, K. and Tompkins, E.L., 2007. Public participation and climate change adaptation: avoiding the illusion of inclusion. *Climate Policy*, 7: 46–59.

Field, R.C., 1997. Risk and justice: capitalist production and the environment. *Capitalism, Nature, Socialism*, 8(2): 69–94.

Fielding, J. and Burningham, K., 2005. Environmental inequality and flood hazard. *Local Environment*, 10(4): 379–410.

Fisher, B.S. and Nasar, J.L., 1992. Fear of crime in relation to three exterior site features: prospect, refuge, and escape. *Environment and Behaviour*, 24: 35–65.

Floyd, M.F. and Johnson, C.Y., 2002. Coming to terms with environmental justice in outdoor recreation: a conceptual discussion with research implications. *Leisure Sciences*, 24(1): 59–77.

Food and Fairness Inquiry, 2010. *Food Justice: The Report of the Food and Fairness Inquiry*. Brighton: Food Ethics Council.

Foreman, C.H., 1998. *The Promise and Perils of Environmental Justice*. Washington, DC: Brookings Institution Press.

Forsyth, T., 2003. *Critical Political Ecology: The Politics of Environmental Science*. London: Routledge.

Fothergill, A. and Peek, L.A., 2004. Poverty and disasters in the United States: a review of recent sociological findings. *Natural Hazards*, 32(1): 89–110.

Fothergill, A., Maestas, E. and DeRouen Darlington, J., 1999. Race, ethnicity and disasters in the United States: a review of the literature. *Disasters*, 23(2): 156–73.

Fouillet, A., Rey, G., Laurent, F., Pavillon, G., Bellec, S., Guihenneuc-Jouyaux, C., Clavel, J., Jougla, E. and Hemon, D., 2006. Excess mortality related to the August 2003 heat wave in France. *International Archives of Occupational and Environmental Health*, 80: 16–24.

Foxman, B., Camargo Jr, C.A., Lilienfeld, D., Mays, V., McKeown, R., Ness, R. and Rothenberg, R., 2006. Looking back at Hurricane Katrina: lessons for 2006 and beyond. *AEP*, 16(8): 652–3.

Frank, L., Glanz, K., McCarron, M., Sallis, J., Saelens, B. and Chapman, J., 2006. The spatial distribution of food outlet type and quality around schools in differing built environment and demographic contexts. *Berkeley Planning Journal*, 19: 79–95.

Fraser, N., 1997. *Justice Interruptus: Critical Reflections on the 'Postsocialist' Condition*. New York: Routledge.

Fraser, N., 1999. 'Social justice in an age of identity politics: redistribution, recognition and participation', in Ray, L. and Sayer, A. (eds), *Culture and Economy after the Cultural Turn*. London: Sage, 25–52.

Freeman III, A.M., 1972. The distribution of environmental quality, in Kneese, A.V. and Bower, B.T. (eds), *Environmental Quality Analysis*. Baltimore, MD: Johns Hopkins Press for Resources for the Future.

Freudenburg, W.R., Gramling, R., Laska, S. and Erikson, K.T., 2009. Disproportionality and disaster: Hurricane Katrina and the Mississippi River–Gulf outlet. *Social Science Quarterly*, 90(3): 497–515.

Frickel, S. and Vincent, M.B., 2007. Hurricane Katrina, contamination, and the unintended organization of ignorance. *Technology in Society*, 29: 181–8.

Friedman, D., 1989. The 'environment racism' hoax. *American Enterprise*, 9(6): 75–8.

Friends of the Earth, 2000. *Pollution Injustice*. London: Friends of the Earth (England and Wales).

Friends of the Earth, 2001. *Pollution and Poverty: Breaking the Link*. London: Friends of the Earth (England and Wales).

Friends of the Earth Scotland, 1999. *The Campaign for Environmental Justice*. Edinburgh: Friends of the Earth Scotland.

Frumkin, H., 2001. Beyond toxicity: the greening of environmental health. *American Journal of Preventative Medicine*, 20: 47–53.

Fussel, E., 2006. Leaving New Orleans: Social Stratification, Networks, and Hurricane Evacuation. *Understanding Katrina: Perspectives from the Social Sciences*, http:// understandingkatrina.ssrc.org/Fussell/ (accessed 28/10/10).

Gabe, T., Falk, G., McCarty, M. and Mason, V.M., 2005. *Hurricane Katrina: Social-Demographic Characteristics of Impacted Areas*. Washington, DC: Congressional Research Service Library of Congress.

Galea, S., Tracy, M., Norris, F. and Coffey, S.F., 2008. Financial and social circumstances and the incidence and course of PTSD in Mississippi during the first two years after Hurricane Katrina. *Journal of Traumatic Stress*, 21(4): 357–68.

Gamson, W.A., 1992. *Talking Politics*. New York: Cambridge University Press.

Gardiner, S.M., Caney, S., Jamieson, D. and Shue, H. (eds), 2010. *Climate Ethics: Essential Readings*. Oxford: Oxford University Press.

Geisler, C. and Letsoalo, E., 2000. Rethinking land reform in South Africa: an alternative approach to environmental justice. *Sociological Research Online*, 5(2): U3–U14.

Gelobter, M., 1992. Toward a model of 'environmental discrimination', in Bryant, B. and Mohai, P. (eds), *Race and the Incidence of Environmental Hazards*. Boulder, CO: Westview, 64–81.

Gieryn, T., 1983. Boundary-work and the demarcation of science from non-science: strains and interests in professional ideologies of scientists. *American Sociological Review*, 48: 781–95.

Gieryn, T., 1999. *Cultural Boundaries of Science: Credibility on the Line*. Chicago, IL: University of Chicago Press.

Gilliland, J., Holmes, M., Irwin, J.D. and Tucker, P., 2006. Environmental equity is child's play: mapping public provision of recreation opportunities in urban neighbourhoods. *Vulnerable Children and Youth Studies*, 1(3): 256–68.

Gobster, P., 1999. Urban parks as green walls or green magnets? Interracial relations in neighbourhood boundary parks. *Landscape and Urban Planning*, 41: 43–5.

Godsill, R., Huang, A. and Solomon, G., 2009. Contaminants in the air and soil in New Orleans after the flood: opportunities and limitations for community empowerment, in Bullard, R. and Wright, B. (eds), *Race, Place, and Environmental Justice after Hurricane Katrina: Struggles to Reclaim, Rebuild, and Revitalize New Orleans and the Gulf Coast*. Boulder, CO: Westview, 115–38.

Goldman, B.A., 1994. *Not Just Prosperity: Achieving Sustainability with Environmental Justice*. Washington, DC: National Wildlife Federation.

Goldman, B.A., 1996. What is the future of environmental justice? *Antipode*, 28(2): 122–41.

Goldman, B.A. and Fitton, L., 1994. *Toxic Wastes and Race Revisited: An Update of the 1987 Report on the Racial and Socioeconomic Characteristics of Communities with Hazardous Waste Sites*. Washington, DC: Center for Policy Alternatives.

Gottlieb, R. and Fisher, A., 1996. 'First feed the face': environmental justice and community food security. *Antipode*, 28(2): 193–203.

Gould, K.A., Pellow, D.N. and Schnaiberg, A., 2004. Interrogating the treadmill of production: everything you wanted to know about the treadmill but were afraid to ask. *Organization and Environment*, 17(3): 296–316.

Gouldson, A., 2006. Do firms adopt lower standards in poorer areas? Corporate social responsibility and environmental justice in the EU and the US. *Area*, 38(4): 402–12.

Gowda, M.V.R. and Easterling, D., 2000. Voluntary siting and equity: the MRS facility experience in Native America. *Risk Analysis*, 20(6): 917–29.

Grasso, M., 2009. An ethical approach to the international-level funding of adaptation to climate change. *IARU International Scientific Congress on Climate Change March 10–12, 2009*. Copenhagen.

Green, C., 2004. The evaluation of vulnerability to flooding. *Disaster Prevention and Management*, 13(4): 323.

Grimes, M., 2005. *Democracy's Infrastructure: The Role of Procedural Fairness in Fostering Consent*. Göteborg: Göteborg University.

Grineski, S.E., 2006. Local struggles for environmental justice: activating knowledge for change. *Journal of Poverty*, 10(3): 25–49.

Grineski, S.E., 2007. Incorporating health outcomes into environmental justice research: the case of children's asthma and air pollution in Phoenix, Arizona. *Environmental Hazards*, 7: 360–71.

Grineski, S.E., Bolin, B. and Boone, C., 2007. Criteria air pollution and marginalized populations: environmental inequity in metropolitan Phoenix, Arizona. *Social Science Quarterly*, 88(2): 535–54.

Groenewegen, P., van den Berg, A.E. and de Vries, S., 2006. Vitamin G: effects of greenspace on health, well being and social safety. *British Medical Journal Public Health*, 6: 149.

Gunewardena, N., Schuller, M. and De Waal, A. (eds), 2008. *Capitalizing on Catastrophe: Neoliberal Strategies in Disaster Reconstruction*. Lanham, MD: AltaMira Press.

Handmer, J.W., Dovers, S. and Downing, T.E., 1999. Societal vulnerability to climate change and variability. *Mitigation and Adaptation Strategies for Global Change*, 4: 267–81.

Handy, F., 1977. Income and air pollution in Hamilton, Ontario. *Alternatives*, 6: 18–24.

Harvey, D., 1973. *Social Justice and the City*. Baltimore, MD: Johns Hopkins Press.

Harvey, D., 1996. *Justice, Nature and the Geography of Difference*. Oxford: Blackwell.

Heiman, M.K., 1996. Waste management and risk assessment: environmental discrimination through regulation. *Urban Studies*, 17(5): 400–18.

Henry, J., 2010. Continuity, social change and Katrina. *Disasters*, 35(1): 220–42.

Hensley, L. and Varela, R.E., 2008. PTSD symptoms and somatic complaints following Hurricane Katrina: the roles of trait anxiety and anxiety sensitivity. *Journal of Clinical Child and Adolescent Psychology*, 37(3): 542–52.

Herzog, T.R. and Chernick, K.K., 2000. Tranquility and danger in urban and natural settings. *Journal of Environmental Psychology*, 20: 29–39.

Heynen, N., 2003. The scalar production of injustice within the urban forest. *Antipode*, 35(5): 980–98.

Heynen, N., 2006a. Green urban political ecologies: toward a better understanding of inner-city environmental change. *Environment and Planning A*, 38: 499–516.

Heynen, N., 2006b. 'The justice of eating in the city: the political ecology of urban hunger', in Heynen, N., Kaika, M. and Swyngedouw, E. (eds), *In the Nature of Cities: Urban Political Ecology and the Politics of Urban Metabolism.* London: Routledge, 129–42.

Heynen, N., Kaika, M. and Swyngedouw, E. (eds), 2006. *In the Nature of Cities: Urban Political Ecology and the Politics of Urban Metabolism.* London: Routledge.

Heynen, N., McCarthy, J., Prudham, S. and Robbins, P. (eds), 2007. *Neoliberal Environments: False Promises and Unnatural Consequences.* New York: Routledge.

Higgs, G. and Langford, M., 2009. GIS science, environmental justice and estimating populations at risk: the case of landfills in Wales. *Applied Geography*, 29: 63–76.

Hillman, M., 2004. The importance of environmental justice in stream rehabilitation. *Ethics, Place and Environment*, 7(1–2): 19.

Hillman, M., 2005. Justice in river management: community perceptions from the Hunter Valley, New South Wales, Australia. *Geographical Research*, 43(2): 152.

Hillman, M., 2006. Situated justice in environmental decision-making: lessons from river management in Southeastern Australia. *Geoforum*, 37(5): 695–707.

Hoerner, J.A. and Robinson, N., 2008. *A Climate of Change: African Americans, Global Warming, and a Just Climate Policy for the U.S.* Oakland, CA: Environmental Justice and Climate Change Initiative.

Hoffman, S.M., 2001. Negotiating eternity: energy policy, environmental justice and the politics of nuclear waste. *Bulletin of Science, Technology and Society*, 21: 456–72.

Holifield, R., 2001. Defining environmental justice and environmental racism. *Urban Geography*, 22(1): 78–90.

Holifield, R., 2004. Neoliberalism and environmental justice in the United States Environmental Protection Agency: translating policy into managerial practice in hazardous waste remediation. *Geoforum*, 35(3): 285–97.

Holifield, R., 2009. Actor-network theory as a critical approach to environmental justice: a case against synthesis with urban political ecology. *Antipode*, 41(4): 637–58.

Holifield, R., Porter, M. and Walker, G. (eds), 2010. *Spaces of Environmental Justice.* Oxford: Wiley.

Honneth, A., 1995. *The Struggle for Recognition: The Moral Grammar of Social Conflicts.* Cambridge, MA: MIT Press.

Hornberg, C. and Pauli, A., 2007. Child poverty and environmental justice. *International Journal of Environmental Health*, 210: 571–80.

Huang, C. and Hwang, R., 2009. 'Environmental justices': what have we learned from the Taiwanese environmental justice controversy. *Environmental Justice*, 2(3): 101–8.

Hunold, C. and Young, I.M., 1998. Justice, democracy and hazardous siting. *Political Studies*, 46(1): 82–95.

Hunt, J.S., Armenta, B.E., Seifert, A.L. and Snowden, J. 2009. The other side of the diaspora: race, threat, and the social psychology of evacuee reception in predominantly white communities. *Organization and Environment*, 22(4): 437–47.

Huq, S., 2001. Climate change and Bangladesh. *Science*, 294: 1617.

Huq, S. and Khan, M.R., 2006. Equity in national adaptation programs of action: the case of Bangladesh, in Adger, N., Paavola, J., Huq, S. and Mace, M.J. (eds), *Fairness in Adaptation to Climate Change.* Cambridge, MA: MIT Press, 181–200.

Hurley, A., 1995. The social biases of environmental change in Gary, Indiana, 1945–80. *Environmental Review*, 12(4): 1–19.

Hurley, A., 1997. Fiasco at Wagner Electric: environmental justice and urban geography in St Louis. *Environmental History*, 2(4): 460–81.

Ikeme, J., 2003. Equity, environmental justice and sustainability: incomplete approaches in climate change politics. *Global Environmental Change: Human and Policy Dimensions*, 13: 195–206.

Iles, A., 2004. Mapping environmental justice in technology flows: computer waste impacts in Asia. *Global Environmental Politics*, 4(4): 76–107.

Illsley, B.M., 2002. Good neighbour agreements: the first step to environmental justice? *Local Environment*, 7(1): 69–79.

Indymedia, 2004. *Dump Protesters Take Direct Action against Landfill Company*, http://www.indymediascotland.org/node/856 (accessed 18/3/2011).

International Federation of Red Cross and Red Crescent Societies, 2007. *World Disasters Report: Focus on Discrimination*. Bloomfield, CT: Kumarian Press.

Irwin, A., Simmons, P. and Walker, G.P., 1999. Faulty environments and risk reasoning: the local understanding of industrial hazards. *Environment and Planning A*, 31: 1311–26.

Ishiyama, N., 2003. Environmental justice and American Indian tribal sovereignty: case study of a land-use conflict in Skull Valley, Utah. *Antipode*, 35(1): 119–39.

Jackson, T. and Illsley, B., 2007. An analysis of the theoretical rationale for using strategic environmental assessment to deliver environmental justice in the light of the Scottish Environmental Assessment Act. *Environmental Impact Assessment Review*, 27(7): 607–23.

Jacobs, M., 1999. *Environmental Modernisation: The New Labour Agenda*. London: Fabian Society.

James, X., Hawkins, A. and Rowel, R., 2007. An assessment of the cultural appropriateness of emergency preparedness communication for low income minorities. *Journal of Homeland Security and Emergency Management*, 4(3): 1–24.

Jamison, A., 2001. *The Making of Green Knowledge*. Cambridge: Cambridge University Press.

Jeffreys, K., 1994. Environmental racism: a skeptic's view. *St John's Journal of Legal Commentary*, 9(2): 677–91.

Jerrett, M., Eyles, J., Cole, D. and Reader, S., 1997. Environmental equity in Canada: an empirical investigation into the income distribution of pollution in Ontario. *Environment and Planning A*, 29(10): 1777–1800.

Jerrett, M., Burnett, R.T., Kanaroglou, P., Eyles, J., Finkelstein, N., Giovis, C. and Brook, J.R., 2001. A GIS environmental justice analysis of particulate air pollution in Hamilton, Canada. *Environment and Planning A*, 33(6): 955–73.

Joassart-Marcelli, P., 2010. Leveling the playing field? Urban disparities in funding for local parks and recreation in the Los Angeles region. *Environment and Planning A*, 42: 1174–92.

Johnson, C., Penning-Rowsell, E. and Parker, D., 2007. Natural and imposed injustices: the challenges in implementing 'fair' flood risk management policy in England. *Geographical Journal*, 173(4): 374–90.

Jorgensen, A., Hitchmough, J. and Calvert, T., 2002. Woodland spaces and edges: their impact on perception of safety and preference. *Landscape and Urban Planning*, 60: 135–50.

Jorgensen, A., Hitchmough, J. and Dunnett, N., 2007. Woodland as a setting for housing-appreciation and fear and the contribution to residential satisfaction and place identity in Warrington New Town, UK. *Landscape and Urban Planning*, 79: 273–87.

Kalan, H. and Peek, B., 2005. South African perspectives on transnational environmental justice networks, in Pellow, D.N. and Brulle, R.J. (eds), *Power, Justice and the Environment: A Critical Appraisal of the Environmental Justice Movement.* Cambridge, MA: MIT Press, 253–63.

Kates, R. and Haarmann, V., 1992. Where the poor live: are the assumptions correct? *Environment*, 34: 4–28.

Katz, C., 2008. Bad elements: Katrina and the scoured landscape of social reproduction. *Gender Place and Culture*, 15(1): 15–29.

Keil, R., 2003. Urban political ecology. *Urban Geography*, 24(8): 723–38.

Kelman, I., 2010. Hearing local voices from small island developing states for climate change. *Local Environment*, 15(7): 605–20.

Kerkof, A.C., Benders, R.M.J. and Moll, H.C., 2009. Determinants of variation in household CO_2 emissions between and within countries. *Energy Policy*, 37: 1509–17.

Ketteridge, A. and Fordham, M., 1998. Flood evacuation in two communities in Scotland: lessons from European research. *International Journal of Mass Emergencies and Disasters*, 16(2): 119–43.

King, D., 2001. Uses and limitations of socio-economic indicators of community vulnerability to natural hazards: data and disasters in Northern Australia. *Natural Hazards*, 24: 147–56.

King, R., 2009. Post-Katrina profiteering, in Bullard, R.D. and Wright, B. (eds), *Race, Place and Environmental Justice after Katrina.* Boulder, CO: Westview, 169–82.

Korc, M.E., 1996. A socioeconomic assessment of human exposure to ozone in the south coast air basin of California. *Journal of the Air and Waste Management Association*, 46(6): 547–57.

Krakoff, S., 2002. Tribal sovereignty and environmental justice, in Mutz, K., Bryner, G.C. and Kenney, D.S. (eds), *Justice and Natural Resources: Concepts, Strategies and Applications.* Washington, DC: Island Press, 161–86.

Krieg, E.J. and Faber, D.R., 2004. Not so black and white: environmental justice and cumulative impact assessments. *Environmental Impact Assessment Review*, 24(7–8): 667–94.

Kruize, H., Driessen, P.J., Glasbergen, P., Van Egmond, K. and Dassen, T., 2007. Environmental equity in the vicinity of Amsterdam airport: the interplay between market forces and government policy. *Journal of Environmental Planning and Management*, 50(6): 699–726.

Kuehn, R.R., 1996. The environmental justice implications of quantitative risk assessment. *University of Illinois Law Review*, 1: 103–72.

Kurtz, H.E., 2002. The politics of environmental justice as the politics of scale: St James Parish, Louisiana and the Shintech Siting Controversy, in Herod, A. and Wright, M. (eds), *Geographies of Power: Placing Scale.* Oxford: Blackwell, 249–73.

Kurtz, H.E., 2007. Gender and environmental justice in Louisiana: blurring the boundaries of public and private spheres. *Gender, Place and Culture*, 14(4): 409–26.

Kurtz, H.E., 2009. Acknowledging the racial state: an agenda for environmental justice research. *Antipode*, 41(4): 684–704.

Lake, A., Townshend, T.G. and Alvanides, S. (eds), 2010. *Obesogenic Environments*. Wiley-Blackwell.

Lake, R.W., 1995. Volunteers, NIMBYs, and environmental justice: dilemmas of democratic practice. *91st Annual Meeting of the Association-of-American-Geographers*, Chicago, IL, 160.

Landry, S.M. and Chakraborty, J., 2009. Street trees and equity: evaluating the spatial distribution of an urban amenity. *Environment and Planning A*, 41: 2651–70.

Lejano, R.P., Piazza, B. and Houston, D., 2002. Rationality as social justice and the spatial-distributional analysis of risk. *Environment and Planning C: Government and Policy*, 20(6): 871–88.

Leonard, L. and Pelling, M., 2010. Mobilisation and protest: environmental justice in Durban, South Africa. *Local Environment*, 15(2): 137–51.

Lepawsky, J. and McNabb, C., 2010. Mapping international flows of electronic waste. *Canadian Geographer*, 54(2): 177–95.

Lewis, S. and Henkels, D., 1998. Good neighbor agreements: a tool for social and environmental justice, in Williams, C. (ed.), *Environmental Victims*. London: Earthscan, 125–41.

Lindsey, G., Maraj, M. and Kuan, S., 2010. Access, equity, and urban greenways: an exploratory investigation. *The Professional Geographer*, 53(3): 332–46.

Lipfert, F.W., 2004. Air pollution and poverty: does the sword cut both ways? *Journal of Epidemiology and Community Health*, 58: 2–3.

Liu, F., 1996. Urban ozone plumes and population distribution by income and race: a case study of New York and Philadelphia. *Journal of the Air and Waste Management Association*, 46(3): 207–15.

Liu, F., 1998. Who will be protected by the EPA's new Ozone and Particulate Matter Standards? *Environmental Science and Technology*, 32(1): 32A–39A.

Liu, F., 2001. *Environmental Justice Analysis: Theories, Methods and Practice*. Boca Raton, FL: Lewis Publishers.

Logan, R., 2006. *The Impact of Katrina: Race and Class in Storm-Damaged Neighborhoods. Report by the American Communities Project*. Available at Brown University: www.s4.brown.edu/Katrina/report.pdf (accessed 13/9/10).

Loh, P., Sugerman-Brozan, J., Wiggins, S., Noiles, D. and Archibald, C., 2002. From asthma to AirBeat: community-driven monitoring of fine particles and black carbon in Roxbury, Massachusetts. *Environmental Health Perspectives*, 110: 297–301.

Lohmann, L., 1999. *The Dyson Effect: Carbon 'Offset' Forestry and the Privatization of the Atmosphere*. London: The Corner House, http://www.thecornerhouse.org.uk/resource/dyson-effect-0.

Lopez, R. and Hynes, P., 2006. Obesity, physical activity, and the urban environment: public health research needs. *Environmental Health: A Global Access Science Source*, 5(25): 5–25.

Loukaitou-Sideris, A., 1995. Urban form and social context: cultural differentiation in the uses of urban parks. *Journal of Planning Education and Research*, 14: 89–102.

Low, N. and Gleeson, B., 1998. *Justice, Society and Nature: An Exploration of Political Ecology*. London: Routledge.

Lucas, K., Walker, G., Eames, M., Fay, H. and Poustie, M., 2004. *Environment and Social Justice: Rapid Research and Evidence Review*. London: Sustainable Development Research Network and Policy Studies Institute.

Luke, T., 2008. The politics of true convenience or inconvenient truth: struggles over how to sustain capitalism, democracy, and ecology in the 21st century. *Environment and Planning A*, 40(8): 1811–24.

Lyle, J.M., 2000. Reactions to EPA's Interim Guidance: the growing battle for control over environmental justice decision-making. *Indiana Law Journal*, 75: 687–708.

Maantay, J. and Maroko, A., 2009. Mapping urban risk: flood hazards, race, and environmental justice in New York. *Applied Geography*, 29(1): 111–24.

McCarthy, J. and Prudham, S., 2004. Neoliberal nature and the nature of neoliberalism. *Geoforum*, 35(3): 275–83.

McConnell, J., 2002. First Minister's speech on the Scottish Executive policy on environment and sustainable development to the Dynamic Earth Conference, Edinburgh, 18 February 2002, http://www.scotland.gov.uk/News/News-Extras/57 (accessed 16/08/2005).

McCracken, R. and Jones, G., 2003. The Aarhus Convention. *Journal of Planning and Environment Law*, 7: 802–11.

McDonald, D.A., 2002. *Environmental Justice in South Africa*. Cape Town: University of Cape Town Press.

McDonald, D.A., 2005. Environmental racism and neo-liberal disorder in South Africa, in Bullard, R.D. (ed.), *The Quest for Environmental Justice: Human Rights and the Politics of Pollution*. San Francisco, CA: Sierra Club Books, 255–78.

McEntire, D.A., 2006. Why vulnerability matters: exploring the merit of an inclusive disaster reduction concept. *Disaster Prevention and Management*, 14(2): 206–22.

McLeod, H., Langford, I.H., Jones, A.P., Stedman, J.R., Day, J.R., Lorenzoni, I. and Bateman, I.J., 2000. The relationship between socio-economic indicators and air pollution in England and Wales: implications for environmental justice. *Regional Environmental Change*, 12: 78–85.

Madge, C., 1997. Public parks and the geography of fear. *Tijdschrift Voor Economische Sociale Geographie*, 88: 237–50.

Madrid, P.A. and Grant, R., 2008. Meeting mental health needs following a natural disaster: lessons from Hurricane Katrina. *Professional Psychology: Research and Practice*, 39(1): 86–92.

Maroko, A.R., Maantay, J.A., Sohler, N.L., Grady, K.L. and Arno, P.S., 2009. The complexities of measuring access to parks and physical activity sites in New York City: a quantitative and qualitative approach. *International Journal of Health Geographics*, 8(34), doi: 10.1186/1476-1072X-1188-1134.

Martinez-Allier, J., 2002. *The Environmentalism of the Poor: A Study of Ecological Conflicts and Valuation*. Cheltenham: Edward Elgar.

Masozera, M., Bailey, M. and Kerchner, C., 2007. Distribution of impacts of natural disasters across income groups: a case study of New Orleans. *Ecological Economics*, 63: 299–306.

Meadowcroft, J., 2008. From welfare state to environmental state? *Journal of European Social Policy*, 18(4): 331–4.

Medina-Ramon, M., Zanobetti, A., Cavanagh, D.P. and Schwartz, J., 2006. Extreme temperatures and mortality: assessing effect modification by personal characteristics and specific cause of death in a multi-city case-only analysis. *Environmental Health Perspectives*, 114: 1331–6.

Mennis, J.L., 2005. The distribution and enforcement of air polluting facilities in New Jersey. *Professional Geographer*, 57(3): 411–22.

Meyer, A., 2000. *Contraction and Convergence*. Totnes: Green Books.

Miller, D., 1995. Equality in post-modern times, in Miller, D. and Walzer, M. (eds), *Pluralism, Justice and Equality*. New York: Oxford University Press, 17–44.

Mitchell, G. and Dorling, D., 2003. An environmental justice analysis of British air quality. *Environment and Planning A*, 35(5): 909–29.

Mitchell, G. and Walker, G., 2007. Methodological issues in the assessment of environmental equity and environmental justice, in Deakin, M., Mitchell, G., Vreeker, R. and Nijkamp, P. (eds), *Sustainable Development: The Environmental Assessment Methods*. London: Taylor and Francis, 447–72.

Mohai, P., 1995. The demographics of dumping revisited: examining the impact of alternate methodologies in environmental justice research. *Virginia Environmental Law Journal*, 14: 615–53.

Mohai, P. and Bryant, B., 1992. Environmental racism: reviewing the evidence, in Bryant, B. and Mohai, P. (ed.), *Race and the Incidence of Environmental Hazards: A Time for Discourse*. Boulder, CO: Westview Press, 163–76.

Mohai, P. and Saha, R., 2006. Reassessing racial and socioeconomic disparities in environmental justice research. *Demography*, 43(2): 383–99.

Montague, P., 2004. Deceptive science: the problem with risk assessment. *Race, Poverty, Environment*, XI(2): 20–2.

Moore, P. and Pastahia, F. (eds), 2007. *Environmental Justice and Rural Communities: Studies from India and Nepal*. Bangkok: IUCN.

Morial, M.H., 2009. Foreword, in Bullard, R. and Wright, B. (eds), *Race, Place and Environmental Justice after Hurricane Katrina*. Boulder, CO: Westview.

Morse, R., 2008. *Environmental Justice through the Eye of Hurricane Katrina*. Washington, DC: Joint Center for Political and Economic Studies, Health Policy Institute.

Mortreux, C. and Barnett, J., 2009. Climate change, migration and adaptation in Funafuti, Tuvalu. *Global Environmental Change: Human and Policy Dimensions*, 19: 105–12.

Most, M.T., Sengupta, R. and Burgener, M.A., 2004. Spatial scale and population assignment choices in environmental justice analyses. *Professional Geographer*, 56(4): 574.

Muller, B., 2001. Varieties of distributive justice in climate change: an editorial comment. Climatic change. *Climate Change*, 48: 273–88.

Muller, B., 2009. *International Adaptation Finance: The Need for an Innovative and Strategic Approach*. Oxford: Oxford Institute for Energy Studies.

Mustafa, D., 2005. The production of an urban hazardscape in Pakistan: modernity, vulnerability and the range of choice. *Annals of the Association of American Geographers*, 95(3): 566–86.

Mutz, K., Gary, C.B. and Douglas, S.K. (eds), 2002. *Justice and Natural Resources: Concepts, Strategies, and Applications*. Washington, DC: Island Press.

Napton, M.L. and Day, F.A., 1992. Polluted neighbourhoods in Texas: who lives there? *Environment and Behaviour*, 24(4): 508–26.

Newell, P., 2005. Race, class and the global politics of environmental inequality. *Global Environmental Politics*, 5(3): 70–94.

Newell, P., 2007. Trade and environmental justice in Latin America. *New Political Economy*, 12(2): 237–59.

Noonan, D.S., 2008. Evidence of environmental justice: a critical perspective on the practice of EJ research and lessons for policy design. *Social Science Quarterly*, 89(5): 1153–74.

Oakes, J.M., Anderton, D. and Anderson, A., 1996. A longitudinal analysis of environmental equity in communities with hazardous waste facilities. *Social Science Research*, 25: 125–48.

O'Brien, M., 2008. *A Crisis of Waste: Understanding the Rubbish Society*. New York: Routledge.

Oelofse, C., Scott, D., Oelofse, G. and Houghton, J., 2006. Shifts within ecological modernization in South Africa: deliberation, innovation and institutional opportunities. *Local Environment*, 11(1): 61–78.

Office of the Inspector General, 2004. *EPA Needs to Consistently Implement the Intent of the Executive Order on Environmental Justice*. Report no. 2004-P-00007. Washington, DC: US Environmental Protection Agency.

Okereke, C., 2006. Global environmental sustainability: intragenerational equity and conceptions of justice in multilateral environmental regimes. *Geoforum*, 37(5): 725–38.

O'Neill, J., 2001. Representing people, representing nature, representing the world. *Environment and Planning C*, 19(4): 483–500.

O'Neill, M.S., Jerrett, M., Kawachi, I., Levy, J.I., Cohen, A.J., Gouveia, N., Wilkinson, P., Fletcher, T., Cifuentes, L. and Schwartz, J. 2003. Health, wealth and air pollution: advancing theory and methods. *Environmental Health Perspectives*, 111(16): 1861–70.

O'Neill, M.S., Zanobetti, A. and Schwartz, J., 2005. Disparities by race in heat-related mortality in four US cities: the role of air-conditioning prevalence. *Journal of Urban Health*, 82(2): 191–7.

O'Riordan, T., Preston-Whyte, R. and Manqele, M., 2000. The transition to sustainability: a South African perspective. *South African Geographical Journal*, 82(2): 1–34.

Osti, R., 2004. Forms of community participation and agencies' role for the implementation of water-induced disaster management: protecting and enhancing the poor. *Disaster Prevention and Management*, 13(4): 6–12.

Oxfam International, 2008. *Climate Wrongs and Human Rights: Putting People at the Centre of Climate-Change Policy*, Oxfam Briefing Paper 117. Oxford: Oxfam International.

Paavola, J., Adger, N. and Huq, S., 2006. Multifaceted justice in adaptation to climate change, in Adger, N., Paavola, J. and Huq, S. (eds), *Fairness in Adaptation to Climate Change*. Cambridge, MA: MIT Press, 1–19.

Pastor, M., Sadd, J. and Hipp, J., 2001. Which came first? Toxic facilities, minority move-in, and environmental justice. *Journal of Urban Affairs*, 23(1): 1–21.

Pastor, M., Sadd, J. and Morello-Frosch, R., 2004. Reading, writing and toxics: children's health, academic performance and environmental justice in Los Angeles. *Environment and Planning C*, 22(2): 271–90.

Pastor, M., Morello-Frosch, R. and Sadd, J.L., 2006. Breathless: schools, air toxics, and environmental justice in California. *Policy Studies Journal*, 34(3): 337–62.

Pastor, M., Sadd, J. and Morello-Frosch, R., 2007. *Still Toxic after All These Years: Air Quality and Environmental Justice in the San Francisco Bay Area*. Santa Cruz: Center for Justice, Tolerance and Community, University of California.

Patel, Z., 2006. Of questionable value: the role of practitioners in building sustainable cities. *Geoforum*, 37(5): 682–94.

Patel, Z., 2009. Environmental justice in South Africa: tools and trade-offs. *Social Dynamics*, 35(1): 94–110.

Paterson, M. and Stripple, J., 2010. My space: governing individual's carbon emissions. *Environment and Planning D*, 28: 341–62.

Paton, K. and Fairbairn-Dunlop, P., 2010. Listening to local voices: Tuvaluans respond to climate change. *Local Environment*, 15(7): 687–98.

Pearce, J., Kingham, S. and Zawar-Reza, P., 2006. Every breath you take? Environmental justice and air pollution in Christchurch, New Zealand. *Environment and Planning A*, 38: 919–38.

Pearce, J., Richardson, E.A., Mitchell, R.J. and Shortt, N.K., 2010. Environmental justice and health: the implications of the socio-spatial distribution of multiple environmental deprivation for health inequalities in the United Kingdom. *Transactions of the Institute of British Geographers*, 35(4): 522–39.

Pearce, L., 2003. Disaster management and community planning and public participation: how to achieve sustainable hazard mitigation. *Natural Hazards*, 28: 211–28.

Pedlowski, M.A., Da Silva, V.C., Adell, J.J.C. and Heynen, N.C., 2002. Urban forest and environmental inequality in Campos dos Goytacazes, Rio de Janeiro, Brazil. *Urban Ecosystems*, 6: 9–20.

Pelling, M., 1999. The political ecology of flood hazard in urban Guyana. *Geoforum*, 30(3): 249–61.

Pelling, M., 2001. Natural Disaster?, in Castree, N. and Braun, B. (eds), *Social Nature: Theory, Practice and Politics*. Oxford: Blackwell, 170–88.

Pelling, M., 2003a. *Natural Disasters and Development in a Globalizing World*. London: Routledge.

Pelling, M., 2003b. *The Vulnerability of Cities: Natural Disasters and Social Resilience*. London, UK: Earthscan.

Pellow, D.N., 2000. Environmental inequality formation: toward a theory of environmental injustice. *American Behavioral Scientist*, 43(4): 581–601.

Pellow, D.N., 2002. *Garbage Wars*. Cambridge, MA: MIT Press.

Pellow, D.N., 2006. Transnational alliances and global politics: new geographies of environmental justice struggle, in Heynen, N., Kaika, M. and Swyngedouw, E. (eds), *In the Nature of Cities: Urban Political Ecology and the Politics of Urban Metabolism*. New York: Routledge, 226–44.

Pellow, D.N., 2007. *Resisting Global Toxics: Transnational Movements for Environmental Justice*. Boston, MA: MIT Press.

Pellow, D.N. and Brulle, R.J. (eds), 2005. *Power, Justice and the Environment: A Critical Appraisal of the Environmental Justice Movement*. Boston, MA: MIT Press.

Pellow, D.N., Steger, T. and McLain, R., 2005. *Proceedings from the Transatlantic Initiative to Promote Environmental Justice Workshop*, Central European University, Budapest, Hungary, 27–30 October 2005, http://cepl.ceu.hu/system/files/Final (accessed 28/10/2008).

Perkins, H., 2010. Green spaces of self-interest within shared urban governance. *Geography Compass*, 4(3): 255–68.

Pettit, J., 2004. Climate justice: a new social movement for atmospheric rights. *IDS Bulletin*, 35(3): 102.

Petts, J., 2005. Enhancing environmental equity through decision-making: learning from waste management. *Local Environment*, 10(4): 397–409.

Pfefferbaum, B., Houston, J.B., Wyche, K.F., Van Horn, R.L., Reyes, G., Jeon-Slaughter, H. and North, C.S., 2008. Children displaced by Hurricane Katrina: a focus group study. *Journal of Loss and Trauma*, 13(4): 303–18.

Phillips, C.V. and Sexton, K., 1999. Science and policy implications of defining environmental justice. *Journal of Exposure Analysis and Environmental Epidemiology*, 9(1): 9–17.

Picou, J.S. and Marshall, B.K., 2007. Social impacts of Hurricane Katrina on displaced K-12 students and educational institutions in coastal Alabama counties: some preliminary observations. *Sociological Spectrum*, 27: 767–80.

Poustie, M., 2004. *Environmental Justice in SEPA's Environmental Protection Activities: A Report for the Scottish Environment Protection Agency*. Glasgow: University of Strathclyde Law School.

Preston, I. and White, V., 2010. *Distributional Impacts of UK Climate Change Policies: Final Report to eaga Charitable Trust*. Bristol: Centre for Sustainable Energy and Association for the Conservation of Energy.

Proctor, J., 2001. Solid rock and shifting sands: the moral paradox of saving a socially constructed nature, in Castree, N. and Braun, B. (eds), *Social Nature*. Oxford: Blackwell, 225–40.

Pulido, L., 1994. Restructuring and the contraction and expansion of environmental rights in the United States. *Environment and Planning A*, 26: 915–36.

Pulido, L., 1996. A critical review of the methodology of environmental racism research. *Antipode*, 28(2): 142–59.

Pulido, L., 2000. Rethinking environmental racism: white privilege and urban development in Southern California. *Annals of the Association of American Geographers*, 90(1): 12–40.

Pye, S., Stedman, J.R., Adams, M. and King, K., 2001. *Further Analyses of N02 and PM10 Air Pollution and Social Deprivation*. Culham: National Environmental Technology Centre.

Pyles, L., Kulkarni, S. and Lein, L., 2008. Economic survival strategies and food insecurity: the case of Hurricane Katrina in New Orleans. *Journal of Social Service Research*, 34(3): 43–53.

Rawls, J., 1972. *A Theory of Justice*. Oxford: *Clarendon Press*.

Republic of South Africa, 1996. *The Constitution of the Republic of South Africa*. Cape Town: RSA.

Richardson, E.A., Shortt, N.K. and Mitchell, R.J., 2010. The mechanism behind environmental inequality in Scotland: which came first, the deprivation or the landfill? *Environment and Planning A*, 42: 223–40.

Ringquist, E.J., 1998. A question of justice: equity in environmental litigation, 1974–1991. *Journal of Politics*, 60(4): 1148–65.

Ringquist, E.J., 2005. Assessing evidence of environmental inequities: a meta-analysis. *Journal of Policy Analysis and Management*, 24(2): 223–47.

Rishbeth, C. and Finney, N., 2006. Novelty and nostalgia in urban greenspace: refugee perspectives. *Tijdschrift voor Economische en Sociale Geografie*, 97(3): 281–95.

Roberts, J., 2009. The international dimension of climate justice and the need for international adaptation funding. *Environmental Justice*, 2(4): 185–90.

Roberts, J. and Parks, B.C., 2007. *A Climate of Injustice: Global Inequality, North–South Politics and Climate Policy*. Cambridge, MA: MIT Press.

Robinson, B.H., 2009. E-waste: an assessment of global production and environmental impacts. *Science of the Total Environment*, 408(2): 183–91.

Routledge, P., Nativel, C. and Cumbers, A., 2006. Entangled logics and grassroots imaginaries of global justice networks. *Environmental Politics*, 15(5): 839–59.

Sandweiss, S., 1998. The social construction of environmental justice, in Camacho, D. (ed.), *Environmental Injustices, Political Struggles.* Durham, NC: Duke University Press, 31–58.

Sastry, N., 2009. Displaced New Orleans residents in the aftermath of Hurricane Katrina: results from a pilot survey. *Organization and Environment,* 22(4): 395–409.

Sasz, A. and Meuser, M., 1997. Environmental inequalities: literature review and proposals for new directions in research and theory. *Current Sociology,* 45(3): 99–120.

Sayer, A., 2005. *The Moral Significance of Class.* Cambridge: Cambridge University Press.

Sayer, A., 2009. Geography and global warming: can capitalism be greened? *Area,* 41(3): 350–3.

Scandrett, E., 2007. Environmental justice in Scotland: policy, pedagogy and praxis. *Environmental Research Letters,* 2(4): 045002.

Schlosberg, D., 1999. *Environmental Justice and the New Pluralism.* Oxford: Oxford University Press.

Schlosberg, D., 2004. Reconceiving environmental justice: global movements and political theories. *Environmental Politics,* 13(3): 517–40.

Schlosberg, D., 2007. *Defining Environmental Justice: Theories, Movements and Nature.* Oxford: Oxford University Press.

Schnaiberg, A., 1980. *The Environment: From Surplus to Scarcity.* New York: Oxford University Press.

Schroeder, R., Martin, K.S., Wilson, B. and Sen, D., 2008. Third world environmental justice. *Society and Natural Resources,* 21(7): 547–55.

Schweitzer, L.A. and Stevenson, M.J., 2007. Right answers, wrong questions: environmental justice as urban research. *Urban Studies,* 44: 319–37.

Scottish Executive, 2002. *Meeting the Needs: Priorities, Actions and Targets for Sustainable Development In Scotland.* Paper 2002/14. Edinburgh: Scottish Executive.

Scottish Executive, 2003. *Planning and Open Space, Planning Advice Note 65.* Edinburgh: Scottish Executive.

Sen, A., 1999. *Development as Freedom.* New York: Anchor Books.

Sen, A., 2009. *The Idea of Justice.* London: Allen Lane.

Sexton, K., 1996. Sociodemographic aspects of human susceptibility to toxic chemicals: do class and race matter for realistic risk assessment? *3rd Annual NHEERL Symposium on Susceptibility and Risk Assessment.* Durham, NC, 261–9.

Sexton, K. and Adgate, J.L., 1999. Looking at environmental justice from an environmental health perspective. *Journal of Exposure Analysis and Environmental Epidemiology,* 9(1): 3–8.

Sexton, K., Greaves, I.A., Church, T.R., Adgate, J.L., Ramachandran, G., Tweedie, R.L., Fredrickson, A., Geisser, M., Sikorski, M., Fischer, G., Jones, D. and Ellringer, P., 2000. A school-based strategy to assess children's environmental exposures and related health effects in economically disadvantaged urban neighborhoods. *Journal of Exposure Analysis and Environmental Epidemiology,* 10(6): 682–94.

Sharkey, P., 2007. Survival and death in New Orleans: an empirical look at the human impact of Katrina. *Journal of Black Studies,* 37(4): 482–501.

Shelton, J.E. and Coleman, M.N., 2009. After the storm: how race, class, and immigration concerns influenced beliefs about the Katrina evacuees. *Social Science Quarterly,* 90(3): 480–96.

Sheppard, E., Leitner, H., McMaster, R.B. and Hongguo, T., 1999. GIS-based measures of environmental equity: exploring their sensitivity and significance. *Journal of Exposure Analysis and Environmental Epidemiology*, 9(1): 18–28.

Shibley, M.A. and Prosterman, A., 1998. Silent epidemic, environmental injustice, or exaggerated concern? Competing frames in the media definition of childhood lead poisoning as a public health problem. *Organization Environment*, 11(1): 33–58.

Shmueli, D.F., 2008. Environmental justice in the Israeli context. *Environment and Planning A*, 40: 2384–2401.

Shostak, S., 2004. Environmental justice and genomics: acting on the futures of environmental health. *Science as Culture*, 13(4): 539–61.

Shrader-Frechette, K., 2002. *Environmental Justice: Creating Equality, Reclaiming Democracy*. New York: Oxford University Press.

Shulman, S.W., Katz, J., Quinn, C. and Srivastava, P., 2005. Empowering environmentally-burdened communities in the US: a primer on the emerging role for information technology. *Local Environment*, 10(5): 501–12.

Sicotte, D., 2010. Some more polluted than others: unequal cumulative industrial hazard burdens in the Philadelphia MSA, USA. *Local Environment*, 15(8): 761–74.

Simmons, P. and Walker, G.P., 2005. Technological risk and sense of place: industrial encroachment on place values, in Boholm, A. and Lofstedt, R. (eds), *Facility Siting: Risk, Power and Identify in Land Use Planning*. London: Earthscan.

Simon, T.W., 2000. In defense of risk assessment: a reply to the environmental justice movement's critique. *Human and Ecological Risk Assessment*, 6(4): 555–60.

Sims, R., Medd, W., Mort, M. and Twigger-Ross, C., 2009. When a 'home' becomes a 'house': care and caring in the flood recovery process. *Space and Culture*, 12(3): 303–16.

Smith, J.G. and Johnston, H. (eds), 2002. *Globalization and Resistance: Transnational Dimensions of Social Movements*. Lanham, MD: Rowman and Littlefield.

Smith, K.R. and Mehta, S., 2003. The burden of disease from indoor air pollution in developing countries: comparison of estimates. *International Journal of Hygiene and Environmental Health*, 206: 279–89.

Smith, N., 2006. Foreword, in Heynen, N., Kaika, M. and Swyngedouw, E. (eds), *In the Nature of Cities: Urban Political Ecology and the Politics of Urban Metabolism*. London: Routledge, xi–xv.

Solecki, W.E. and Welch, J.M., 1995. Urban parks: green spaces or green walls? *Landscape and Urban Planning*, 32(2): 93–106.

Souza, A., 2008. The gathering momentum for environmental justice in Brazil. *Environmental Justice*, 1(4): 183–8.

Spatareanu, M., 2007. Searching for pollution havens: the impact of environmental regulations on foreign direct investment. *Journal of Environment and Development*, 16(2): 161–82.

Speller, G., 2005. *Improving Community and Citizen Engagement in Flood Risk Management Decision Making, Delivery and Flood Response*. Bristol: Environment Agency.

Stanley, A., 2009. Just space or spatial justice? Difference, discourse and environmental justice. *Local Environment*, 14(10): 999–1014.

Starkey, R., 2008. *Allocating emissions rights: are equal shares, fair shares?* Tyndall Centre Working Paper 118. University of Manchester, Manchester: Tyndall Centre.

Steger, T. (ed.), 2007. *Making the Case for Environmental Justice in Central and Eastern Europe*. Budapest: CEU Center for Environmental Policy and Law (CEPL), the Health and Environment Alliance (HEAL) and the Coalition for Environmental Justice.

Stein, R., 2004. *New Perspectives on Environmental Justice: Gender, Sexuality and Activism*. Rutgers, NJ: Rutgers University Press.

Steinführer, A., Kuhlicke, C., de Marchi, B., Scolobig, A., Tapsell, S. and Tunstall, S.M., 2007. *Recommendations for flood risk management with communities at risk*: FLOODsite, http://www.floodsite.net/ (accessed 28/10/10).

Stephens, C., Bullock, S. and Scott, A., 2001. *Environmental Justice: Rights and Means to a Healthy Environment for All*. Swindon: ESRC Global Environmental Change Programme.

Stevenson, S., Stephens, C., Landon, M., Pattendon, S., Wilkinson, P. and Fletcher, T., 1998. Examining the inequality and inequity of car ownership and the effects of pollution and health outcomes such as respiratory disease. *Epidemiology* 9(4): S29. Abstract of an oral presentation to the 10th Conference of the International Society for Environmental Epidemiology, Boston.

Stretesky, P. and Hogan, M.J., 1998. Environmental justice: an analysis of Superfund sites in Florida. *Social Problems* 45(2): 268–87.

Strife, S. and Downey, L., 2009. Childhood development and access to nature: a new direction for environmental inequality research. *Organization and Environment*, 22(1): 99–122.

Su, J.G., Morello-Frosch, R., Jesdale, B.M., Kyle, A.D., Shamasunder, B. and Jerrett, M., 2009. An index for assessing demographic inequalities in cumulative environmental hazards with application to Los Angeles, California. *Environmental Science and Technology*, 43: 7626–34.

Susman, P., O'Keefe, P. and Wisner, B., 1983. Global disasters: a radical interpretation, in Hewitt, K. (ed.), *Interpretations of Calamity from the Viewpoint of Human Ecology*. Boston, MA: Allen and Unwin, 263–83.

Sustainable South Bronx, 2008. *In My Backyard: A Profile of Hunts Point with Recommendations for Realizing Community Members' Vision for Their Neighbourhood*. New York: Sustainable South Bronx.

Swyngedouw, E., 2010. Apocalypse forever? Post-political populism and the spectre of climate change. *Theory, Culture and Society*, 27: 213–32.

Sze, J., 2004. Gender, asthma politics and urban environmental justice activism, in Stein, R. (ed.), *New Perspectives on Environmental Justice: Gender, Sexuality and Activism*. New Brunswick, NJ: Rutgers University Press, 177–90.

Sze, J., 2007. *Noxious New York: The Racial Politics of Urban Health and Environmental Justice*. Cambridge, MA: MIT Press.

Sze, J. and London, J.K., 2008. Environmental justice at the crossroads. *Sociology Compass*, 2(4): 1331–54.

Sze, J. and Prakash, S., 2004. Human genetics, environment, and communities of color: ethical and social implications. *Environmental Health Perspectives*, 112(6): 740–5.

Sze, J., Gambirazzio, G., Karner, A., Pastor, M. and Sadd, J., 2009. Best in show? Climate and environmental justice policy in California. *Environmental Justice*, 2(4): 179–84.

Taylor, A.F., Wiley, A., Kuo, F.E. and Sullivan, W.C., 1998. Growing up in the inner city: green spaces as spaces to grow. *Environment and Behaviour*, 30(1): 3–27.

Taylor, A.J. and McGwin, G., 2000. Temperature-related deaths in Alabama. *Southern Medical Journal*, 93(8): 787–92.

Taylor, D.E., 2000. The rise of the environmental justice paradigm: injustice framing and the social construction of environmental discourses. *American Behavioral Scientist*, 43(4): 508–80.

Terry, G., 2009. No climate justice without gender justice: an overview of the issues. *Gender and Development*, 17(1): 5–18.

Tewdwr-Jones, M. and Allmendinger, P., 1998. Deconstructing communicative rationality: a critique of Habermasian collaborative planning. *Environment and Planning A*, 30: 1975–89.

Thompson, D., 2005. Democracy in time: popular sovereignty and temporal representation. *Constellations*, 12(2): 245–61.

Thompson, M., 1979. *Rubbish Theory: The Creation and Destruction of Value*. Oxford: Oxford University Press.

Thrush, D., Burningham, K. and Fielding, J., 2005. *Flood Warning for Vulnerable Groups: A Review of the Literature*. Bristol: Environment Agency.

Timmons-Roberts, J., 2009. The international dimension of climate justice and the need for international adaptation funding. *Environmental Justice*, 2(4): 185–90.

Timpiero, A., Ball, K., Salmon, J., Roberts, R. and Crawford, D., 2007. Is availability of public open space equitable across areas? *Health and Place*, 13(2): 335–40.

Tol, R.S.J., Downing, T.E., Kuik, O.J. and Smith, J.B., 2003. *Distributional Aspects of Climate Change Impacts*, ENV/EPOC/GSP(2003)14/FINAL. Paris: Organisation for Economic Cooperation and Development.

Torres, G., 1994. Environmental burdens and democratic justice. *Fordham Urban Law Journal*, 21: 431–60.

Towers, G., 2000. Applying the political geography of scale: grassroots strategies and environmental justice. *Professional Geographer*, 52(1): 23–36.

Trainor, S.F., Chapin, F.S., Huntington, H.P., Natcher, D.C. and Kofinas, G., 2007. Arctic climate impacts: environmental injustice in Canada and the United States. *Local Environment*, 12(6): 627–43.

Tschakert, P., 2009. Digging deep for justice: a radical re-imagination of the artisanal gold mining sector in Ghana. *Antipode*, 41(4): 706–40.

Twigger-Ross, C. and Scrase, I., 2006. *Developing an Environment Agency Policy on Vulnerability and Flood Incident Management*. Bristol: Environment Agency.

Tzoulas, K. and Korpela, K., 2007. Promoting ecosystem and human health in urban areas using green infrastructure: a literature review. *Landscape and Urban Planning*, 81(3): 167–78.

Ueland, J. and Warf, B., 2006. Racialized topographies: altitude and race in southern cities. *Geographical Review*, 96(1): 50–78.

UK Government, 1999. *A Better Quality of Life: The UK Strategy for Sustainable Development*. London: HMSO.

United Church of Christ, 1987. *Toxic Waste and Race in the United States*. New York: United Church of Christ.

Urban Green Spaces Taskforce, 2002. *Green Spaces, Better Places: Final Report*. London: DTLR.

US Environmental Protection Agency, 2008. *Extract from Environmental Justice Home Page*, http://www.epa.gov/compliance/environmentaljustice/ (accessed 30/10/2008).

US General Accounting Office, 1983. *Siting of Hazardous Waste Landfills and Their Correlation with Racial and Economic Status of Surrounding Communities*. Washington, DC: USGAO.

Vanderheiden, S., 2008. *Atmospheric Justice: A Political Theory of Climate Change*. New York: Oxford University Press.

Vermeylen, S. and Walker, G., 2011. Environmental justice, values and biological diversity: the San and the Hoodia Benefit-Sharing Agreement, in Carmin, J. and

Agyeman, J. (eds), *Environmental Injustice beyond Borders: Local Perspectives on Global Inequities*. Boston, MA: MIT Press, 105–28.

Walker, G., 2007. Environmental justice and the distributional deficit in policy appraisal in the UK. *Environmental Research Letters*, 4: 045004.

Walker, G., 2009a. Beyond distribution and proximity: exploring the multiple spatialities of environmental justice, *Antipode*, 41(4): 614–36.

Walker, G., 2009b. New Nuclear: Environmental Justice and the Politics of Consent. Paper presented at the Nordic Environmental Social Sciences Conference, June 2009, University College London.

Walker, G. and Burningham, K., 2011. Flood risk, vulnerability and environmental justice: evidence and evaluation of inequality in a UK context. *Critical Social Policy*, 31(2): 216–40.

Walker, G. and Shove, E., 2007. Ambivalence, sustainability and the governance of sociotechnical transitions. *Journal of Environmental Policy and Planning*, 9(3/4): 213–25.

Walker, G., Fairburn, J., Smith, G. and Mitchell, G., 2003. *Environmental Quality and Social Deprivation Phase II: National Analysis of Flood Hazard, IPC Industries and Air Quality*. Bristol, UK: Environment Agency.

Walker, G., Mitchell, G., Fairburn, J. and Smith, G., 2005. Industrial pollution and social deprivation: evidence and complexity in evaluating and responding to environmental inequality. *Local Environment*, 10(4): 361–77.

Walker, G., Burningham, K., Fielding, J., Smith, G. and Thrush, D., 2006. *Addressing Environmental Inequalities: Flood Risk*. Bristol: Environment Agency.

Walker, G., Burningham, K., Fielding, J., Smith, G., Thrush, D. and Fay, H., 2007. *Addressing Environmental Inequalities: Flood Risk*, Science Report SC020061. Bristol: Environment Agency.

Walker, G., Damery, S., Smith, G., and Petts, J., 2008. *Addressing Environmental Inequalities: Flood Risk, Waste Management and River Water Quality in Wales*, R&D Technical Report SC020061/SR5. Bristol: Environment Agency.

Walker, M., Whittle, R., Medd, W., Burningham, K., Moran Ellis, J. and Tapsell, S., 2010. *After the Rain: Learning Lessons about Flood Recovery and Resilience from Children and Young People in Hull*. Final project report for 'Children, Flood and Urban Resilience: Understanding Children and Young People's Experience and Agency in the Flood Recovery Process'. Lancaster: Lancaster University.

Walker, G., Whittle, R., Medd, W. and Walker, M., forthcoming. Assembling the flood: producing spaces of bad water in the city of Hull. *Environment and Planning A*, forthcoming.

Walzer, M., 1983. *Spheres of Justice*. New York: Basic Books.

Waterstone, M., 1985. The equity aspects of carbon dioxide-induced climate change. *Geoforum*, 16(3): 301–6.

Watson, M. and Bulkeley, H., 2005. Just waste? Municipal waste management and the politics of environmental justice. *Local Environment*, 10(4): 411–26.

Watts, M. and Peet, R., 2004. Liberating political ecology, in Peet, R. and Watts, M. (eds), *Liberation Ecologies*, 2nd edn. London: Routledge, 3–43.

Weinberg, A.S., 1998. The environmental justice debate: new agendas for a third generation of research. *Society and Natural Resources*, 11(6): 605–14.

Wenz, P.S., 1988. *Environmental Justice*. Albany, NY: Suny Press.

Wenz, P.S., 2000. Environmental justice through improved efficiency. *Environmental Values*, 9(2): 173–88.

Wernette, D.R. and Nieves, L.A., 1992. Breathing polluted air: minorities are disproportionately exposed. *EPA Journal*, 18(2): 16–17.

Werrity, A., Houston, D., Ball, T., Tavendale, A. and Black, A., 2007. *Exploring the Social Impacts of Flood Risk and Flooding in Scotland*. Edinburgh: Scottish Executive.

West, D.M. and Orr, M., 2007. Race, gender, and communications in natural disasters. *Policy Studies Journal*, 35: 569–86.

West Harlem Environmental Action, 2004. *Harlem on the River: Making a Community Vision Real*. New York: WE ACT for Environmental Justice.

West, P.C., 1989. Urban region parks and black minorities: subculture, marginality, and interracial relations in park use in the Detroit metropolitan area. *Leisure Sciences*, 11: 11–28.

Wheeler, B.W., 2004. Health-related environmental indices and environmental equity in England and Wales. *Environment and Planning A*, 36: 802–22.

Wheeler, D., 2000. *Racing to the Bottom? Foreign Investment and Air Quality in Developing Countries*. Washington, DC: World Bank.

Whittle, R., Medd, W., Deeming, H., Kashefi, E., Mort, M., Twigger Ross, C., Walker, G. and Watson, N., 2010. *After the Rain: Learning the Lessons from Flood Recovery in Hull*. Final project report for 'Flood, Vulnerability and Urban Resilience: A Real-Time Study of Local Recovery Following the Floods of June 2007 in Hull'. Lancaster: Lancaster University.

Whyley, C., McCormick, J. and Kempson, E., 1998. *Paying for Peace of Mind: Access to Home Contents Insurance for Low-Income Households*. London: Policies Studies Institute.

Williams, G. and Mawdsley, E., 2006. Postcolonial environmental justice: government and governance in India. *Geoforum*, 37(5): 660–70.

Williams, R.W., 1999. Environmental injustice in America and its politics of scale. *Political Geography*, 18(1): 49–73.

Wilson, J.F., 2006. Health and the environment after Hurricane Katrina. *Annals of Internal Medicine*, 144(2): 153–6.

Wilson, S.M., Richard, R., Joseph, L. and Williams, E., 2010. Climate change, environmental justice and vulnerability: an exploratory spatial analysis. *Environmental Justice*, 3(1): 13–19.

Wisner, B., 2001. 'Vulnerability' in Disaster Theory and Practice: From Soup to Taxonomy, then to Analysis and finally Tool. International Work Conference, Disaster Studies of Wageningen University and Research Centre, 29/30 June 2001.

Wisner, B., Blaikie, P., Cannon, T. and Davis, I., 2004. *At Risk, Natural Hazards, People's Vulnerability and Disasters*. London: Routledge.

Wolch, J., Wilson, J.P. and Fehrenbach, J., 2002. *Parks and Park Funding in Los Angeles: An Equity Mapping Analysis*. Los Angeles: Sustainable Cities Program, GIS Research Laboratory, University of South California, http://www.usc.edu/dept/geography/ESPE (accessed 28/10/10).

Wolsink, M., 2007. Wind power implementation: the nature of public attitudes. Equity and fairness instead of 'backyard motives'. *Renewable and Sustainable Energy Reviews*, 11(6): 1188–1207.

Wrigley, N., 2002. 'Food deserts' in British cities: policy context and research priorities. *Urban Studies*, 39(11): 2029–40.

Yohe, G. and Schlesinger, M., 2002. The economic geography of the impacts of climate change. *Economic Geography*, 2(3): 311–41.

Young, E., 1998. Dealing with hazards and disasters: risk perception and community participation in management. *Australian Journal of Emergency Management*, Winter 1998: 14–16.

Young, I.M., 1990. *Justice and the Politics of Difference*. Princeton, NJ: Princeton University Press.

Young, I.M., 1998. Harvey's complaint with race and gender struggles: a critical response. *Antipode*, 30(1): 36–42.

Young, I.M., 2006. Katrina: too much blame, not enough responsibility. *Dissent*, Winter 2006: 41–6.

Zavestoski, S., Shulman, S. and Schlosberg, D., 2006. Democracy and the environment on the Internet: electronic citizen participation in regulatory rulemaking. *Science Technology and Human Values*, 31(4): 383–408.

Index

References to tables are shown in bold, figures are indicated by references in italics and text within boxes is indicated by the preffix 'b' before the page number.

adaptation policies, carbon emissions 192–8, **195**, 209–10, **210**
adaptive capacity 187–8
Adger, N. and Barnett, J. 197
Adger, N. *et al* 2006 182, 183, 188
African-American community: Hurricane Katrina 141–3; impacts of climate change on 190–1, b191–2; and waste site locations 84–90, b85, b87–8, *88*, 91–4, **91**
age factors: fatalities in Hurricane Katrina 143–4, **143**; susceptibility to heatwaves 191–2, b191–2; vulnerability to flooding 215
Agyeman, J. and Evans, B. 1
Agyeman, J. *et al* 2009 197
air pollution: knowledge of 122–3; metric of distribution **43**; processes and explanations 112–15; responsibility 106–7, **106**, 118–19, 126; studies on patterns of exposure 107–11; traffic 118–19; urban 108–9
air quality: campaigners on air quality (photo) *104*; claim-making 5, 105–7, **106**, 118–19; cross-cutting themes b105; distributional studies 116–17; distributive justice 120–3; evidence collection 123–5; minimum standards 120–3, 126; and poverty 106–7, **106**, 109, 111; rural areas 112–13; statistical methodologies 107–8, **108**, 109–10, **110**, 117–18; studies, USA 109–11, **110**, 117–8; urban areas 5, 112–13, *114*; vulnerability 106–7, **106**, 115–18, 126
Akaba, A. 123

Alaska 196, 197
Alternatives for Community and Environment 122
Anderton, D. *et al* 1994 87
Arctic communities 196, 197
Association of British Insurers 136
asthma, childhood 117, 122, 123–5, *124*
Australia b45

Bangladesh 139, 188, b189–90
Basel Action Group 97
Basel Convention 96
Beckman, L. and Page, E. 182, 183
Been, V. 89
Been, V. and Gupta, F. 89
Bell, D. 42–4
Bell, S. *et al* 2008 160
Bickerstaff, K. and Agyeman, J. 98
Bill of Rights of the South African Constitution 31–2
blame and responsibility 21–2
Block, W. 149
Bowen, W.M. 57–9
Brand, C. and Boardman, B. 205–6, *205*
Brazil 33–4
Briggs, D. *et al* 2008 117
Brown, S. and Walker, G. b191–2
Brownlow, A. 62, 161–2
Bryant, B. b8–9, 10, 51
bucket brigades 125
Bullard, B. 25, 67, 86, 89, 98
Bullard, B. *et al* 2007 b87–8
Bush, George W. 70, 197, 201
Buzzelli, M. 116
Byrne, J. and Wolch, J. *162*

CAG Consultants 190
California Waste Management Board 86
Canada: Arctic communities 196, 197;
    greenspaces 171; Just Earth 179
capabilities framework 51–3, 151–3
Capacity Global 28
capitalism 68–9, 70, 93, 186
carbon consumption 185–6
carbon emissions: adaptation policies
    193–5, 209–10, **210**; carbon dioxide
    203; integration of mitigation and
    adaptation policies 209–10, **210**;
    mitigation policies, global 198–204;
    mitigation policies, national 204–6;
    mitigation policies, social effects 207–9,
    208; offset markets 207; per capita
    emissions, selected countries *199*; total
    emissions, selected countries *200*
carbon intensity 201–2, **201**
carbon trading 206
causal responsibility b45, 46–7, 118–19
Chavis, Benjamin 66
children: access to greenspaces 161–2,
    b164, 170, 171; asthma 117, 122,
    123–5, *124*; effects of Hurricane Katrina
    145–6; poverty rates, New Orleans *141*;
    vulnerability 4, 46
China b181
civil rights movement (USA) 18, 20,
    28, 31
claim-making: defined 5–8, 14; evidence
    40–2; examples of b6; main elements
    40–1, *40*
Clean Air Coalition, South Bronx
    USA 123
Clean Development Mechanism
    (CDM) 207
climate change: adaptation policies 192–8,
    195, 209–10, **210**; adaptive capacity
    187–8; community of justice 42–3, 192–
    5; Environmental Justice and Climate
    Change 10 policies b211–12; impacts
    of 186–8, b189–90, 198; international
    negotiations 180, b181; metric of dis-
    tribution **43**; mitigation policies, global
    198–204; mitigation policies, integration
    with adaptation policies 209–10, **210**;
    mitigation policies, national 204–6;
    mitigation policies, social effects 207–9;
    scales of analysis 35–6, 215–16
climate justice: campaigning 180–2;
    cross-cutting themes b182; framing
    179–80, 202–3; human rights

policies **210**; knowledge claims 182–4,
    204; policy integration 209–12; power
    relations 185–6; procedural justice
    195–8; protest march (photo) *180*;
    scales of analysis 184–5
Clinton, William Jefferson "Bill" 18, b19
Coalition for Environmental Justice
    (in Central and Eastern Europe) b8–9,
    10, 24, 35
Collins, T.W. 134–5
Columbia University, School of Public
    Health 124–5, *124*
Commonwealth of Massachusetts b8–9, 10
Communities against Toxics 25
Communities for a Better
    Environment 123
Community Lobby Opposing Unhealthy
    Tips 25
community of justice 42–3, 48, 123–5,
    192–5
community participation: in decision-
    making process 98–100; in evidence
    generation 55, 123–5; Good Neighbour
    Agreements 125; urban greenspaces
    159, 174–5, 176–7
community-scale compensation 175–6
contextual process claims 64, 219–20
contraction and convergence model
    202, *203*
Copenhagen Summit (2009 United Nations
    Climate Change Conference) b181, 204
Coughlin, S.S. 117
cultural recognition *see* recognition,
    cultural
Cutter, S. and Solecki, W.D. 114

decision-making processes 48–50
Department of Environment, Food and
    Rural Affairs (Defra) 28
developing world: data gathering 56;
    flooding 135, b189–90; urban air
    pollution 108–9
discrimination *see* recognition
discriminatory intent 88–90
distributional studies, air quality 116–17
distributive justice: air quality **43**, 120–3;
    carbon mitigation measures **208**;
    claim-making 42–7; climate justice
    192–4; definition b10; flooding 150–1;
    frame, USA 21; intergenerational 193;
    lack of in South Africa 31; stream
    rehabilitation, Australia b45; urban
    greenspaces 176–7

Dow, K. *et al* 2006 188, 194, **195**
'dumping in Dixie' hypothesis 78–9, 86, 114

Earthlife International Conference 30
ecological debt 185, 203
economic analyses, hazardous waste flows 95–6
Enarson, E. 153–4
Endres, D. 100
England: air quality 112–13, 120–1; greenspace access 173–4; Hull, floods 136, b137–8, *137*; nitrogen dioxide levels *112*, *121*; triple jeopardy argument 116–17; waste site locations b93–4
English Nature 173–4
Environment Agency (EA): analysis of industrial pollution and income 54–5; and environmental justice 26–7; flooding 127, 133; position statement (2004) b27
environmental boundaries 20–1
environmental concerns 1–2, **2**, 3
environmental discrimination 66–8
environmental inequality: definition 12–13; Environment Agency (EA) agenda 26–7; linked with normative positions 5–6, 40; of preparation for floods 129; relationship with injustice 217–18
Environmental Justice and Climate Change (EJCC) b211–12
Environmental Justice and Climate Change Initiative 179
environmental justice movement: definitions 1–2, 8–12, b8–9, 13, 14, 16; as dynamic process 220–1; fields of study 2–3, **2**; frames 4–5, 33; globalisation 23–5; language 1, 2, 17, 18, 23–4, **24**; links with political ecology 72; in South Africa 1, 30–2, 33; in the UK 25–30, 33, 37; in the USA 2, 17–23, 34
Environmental Justice Networking Forum (EJNF) 30, 32
Environmental Justice Summits 28
Environmental Protection Agency (EPA) 70, 79, 176
environmental racism: discriminatory intent 88–90, *88*; Hurricane Katrina 139; institutional **91**, 92, 147; intentional 91–2, **91**; market dynamics **91**, 92–3;

South Africa 31, 33, 67; Taiwan 81; United States of America 2, 18, 66–7, 86, 93–4, 109–11
epidemiology 61, 62, 117, 118, 122
Eskimo people 197
Europe, environmental justice movement 24, **24**
evidence: claim-making 40–2; community participation in 55, 123–5; cumulative risk assessments 59–61, b60; distributional studies 116–17; forms of 53–4; of inequalities 53, 89–90; longitudinal studies 89–90, **90**; meta analysis 90; narrative 62–3; politics of 63, 90; production 54–5, 216–17; proxy data 56–7, 183; qualitative methods 61–3; quantitative analytical methods 55–61, 84, 88, 107–8; research designs b58–9
e-waste 97–8
Executive Order 12898 (USA) 18, b19

Faber, D. 33, 63–4, 70, 74, 101
Fairburn, J. *et al* 2005 *167*, *168*
Fairmount Park, Philadelphia 161–2
flooding *see also* Hurricane Katrina: Bangladesh b189–90; characteristics 129–30; and climate change 184; cross-cutting themes b128–9; distributive justice 150–1; inequalities of population exposure to 130–5; insurance 136, 146; metric of distribution **43**; participatory justice 153–4; as part of environmental justice framework 127; poverty 136–9; pre-flood approach 131; proxy data 57; resilience to 151–3; rivers 133, *133*, 134, 135; role of the state 149, 150–1, 152, 153; sea 133, 134, *134*; temporal characteristics 129–30; toxic contamination 146; vulnerability to 127–8, 135–9, *136*, 147, 149
food justice 159–60
Foreman, C.H. 61
Foxman, B. *et al* 2006 144
framing: characteristics of US frame 20–3, b21–2; countries using an environmental justice frame **24**; defined 4–5, 14, 16, 17–18; first time used in national contexts 82; global form 24–5, 34–7, 218; reframing 26–7, 28, 36, 37; in South Africa 30–3; in the UK 25–30, b27, 32–3

Friends of the Earth (FoE) 25–6, 54, 98
Friends of the Earth International
    (FoEI) 203
Friends of the Earth Scotland (FoES)
    b8–9, 10, 28, 34, 80
Friends of the Earth Scotland (FoES)
    (photo) *29*

Gabe, T. *et al* 2005 *141*
Gary, Indiana 114, b115
gender: misrecognition of evidence 62, 63,
    68; susceptibility to heatwaves 191–2,
    b191–2
General Accounting Office (GAO) 85,
    b85, 89
genocide 195–6
Geographical Information Systems (GIS)
    57, **58**
Gilliland J. *et al* 2006 171, 174
Glasgow, proximity to greenspaces **165**,
    168–9
Global Commons Institute *203*
globalisation (environmental justice)
    16–17, *17*, 23–5, 33, 34–7, 70
Global North and South 36, 184, 186
global warming 187; *see also* climate
    change
Good Neighbour Agreements 125
Gould, K. A. *et al* 2004 69
government policies: decision-making
    processes 48; in Scotland 28–30; in
    South Africa 31–2, 33; in the UK 25,
    26–8, b27, 33; in USA 18–20, b19, 23
grandfathering 200–1, **201**
grassroots civil rights activism:
    involvement in procedural justice
    123–5; lack of in UK 25, 26, 28; South
    Africa 30, 33; United States of America
    18, 20, 23, 28
Greengairs, Scotland 79–80, **83**, 84, 95
greenhouse gases (GHG) *see also* carbon
    emissions 183, 185, 198, 200–4,
    201, *205*
greenspaces *see also* urban greenspaces:
    within an environmental justice
    framework 156; in justice terms 173;
    metric of distribution **43**; minimum
    standards 173–4; narrative evidence
    62; procedural justice 176–7; types of
    greenspaces 165–6, **169**; woodlands,
    Scotland 166–7, *167*, *168*
Grineski, S.E. 117–18
groundWork 30

Handmer, J.W. *et al* 1999 188
Harvey, D. 98
hazardous waste, exportation of 70, 95–8
hazardous waste landfill sites 79, 82
health benefits, urban greenspaces 158
heatwaves 191–2, b191–2
Heynen, N. 163
Heynen, N. *et al* 2006 72–4
Hillman, M. 45, b45, 46
historical responsibility, greenhouse gas
    emissions **201**, 203
Hoerner, J.A. and Robinson, N. 191
Hoffman, S.M. 99–100
Houston, City of 89
Hull, Yorkshire (Kingston upon Hull) 136,
    b137–8, *137*
Hunold, C. and Young, I.M. 49, **49**
Hurley, A. 114, b115
Hurricane Katrina: concepts of justice
    147–8, **148**; evacuation 143; fatality
    rates 143–4, **143**, 146; as focus for
    environmental activism 127, 139–41;
    Katrina Bill of Rights b152; New
    Orleans under water (photo) *140*;
    socio-demographic profile those
    affected 141–3, *141*, *142*, 185

Indianapolis, Indiana 171–2
inequality *see* environmental inequality
institutional racism **91**, 92, 147
insurance, flood 136, 146
intergenerational distributional justice 193
international negotiations, climate change
    180, b181
'is-ought' distinction 12–13, 40
Israel 33
Ivory Coast b96–7

Jeffreys, K. 92, 93
justice *see* distributive justice; procedural
    justice; recognition, justice as

Kalan, H. and Peek, B. 30
Katrina Bill of Rights b152
Katz, C. 145, 146–7
Kenya 139
Krakoff, S. 100
Kyoto Protocol 200–1

Labour Party, 'New Labour' 25
Lake, R.W. 49–50, 98, 101
landfill sites: Greengairs, Scotland 79–80;
    Ivory Coast b96–7

Landry, S.M. and Chakraborty, J. 172
language of environmental justice 1, 2, 17,
    18, 23–4, **24**
Lepawsky, J. and McNabb, C. 97–8
libertarian perspectives: post-Hurricane
    Katrina 149, 219; urban greenspaces
    173, 187
Lindsey, G. *et al* 2010 171–2
Little Village Environmental Justice
    Organization 21, b21–2, *22*
lock-in, carbon dependency 185–6
Logan, R. *142*
Lohmann, L. 207
longitudinal studies 89–90, **90**
Los Angeles, USA 170
low carbon transition *see* mitigation
    policies

Maantay, J. and Maroko, A. 131
market dynamics hypothesis 89, 92–3
Maroko, A.R. *et al* 2009 171
Martinez-Allier, J. 31, 34, 36
McConnell, Jack 28–9
Meadowcroft, J. 207
methodologies, statistical: air quality
    107–8, **108**, 109–10, **110**, 117–18;
    analysing greenspaces 165–6, 170–2;
    complexities of 216–17; Geographical
    Information Systems (GIS) **58**; waste
    site locations 85–8
metrics of distributive justice 43–4, **43**
minimum standards: air quality 120–3,
    126; greenspaces 173–4
misrecognition *see also* recognition,
    justice as: on basis of gender 62, 63, 68;
    disability 51
mitigation policies, carbon emissions:
    global 198–204; integration with
    adaptation policies 209–10, **210**;
    national 204–6; social effects 207–9,
    208
Morial, M.H. b152–3
Morse, R. 142
Movement Generation Justice and Ecology
    Project b181

narrative evidence 62–3
national sustainable development strategy
    (UK Government 1999) 28
National Urban League 152
nation blocks 184, 199
Native Americans 99–100
Netherlands 114

Newell, P. 35
New Orleans, Louisiana *see also*
    Hurricane Katrina: flooded (photo) *140*;
    reconstruction 146–7; relocation and
    return to 144–5; social impacts after the
    flood 145–6
New York, USA 171, 174–5, *175*, 214
New Zealand 119
nitrogen dioxide: around Amsterdam
    airport 114; car pollution 118–19;
    levels in the UK 111, 112–13, *112*,
    120–1; levels in Wales *113*, 120;
    nitrogen dioxide levels *121*
Non-governmental organizations (NGOs):
    in the UK 25, 26, 28, 33; worldwide 32,
    96, 153
Noonan, D. 92–3
normative positions: greenspaces 176;
    injustice 14, 17; linked with inequalities
    5–6; reasoning 40, 41
North Korea 81
nuclear waste 101
nuclear waste facility, Orchid Island,
    Taiwan 80–4, *81*, **83**, 99

Obama, Barack Hussein 20, b181
obesogenic environments 158
Office of Environmental Justice (USA) 18
Office of Inspector General (USA) b19
oil refineries 30
Orchid Island (Lanyu), Taiwan 80–4, *81*,
    83, 99
Oxfam International 196, 197, 210
ozone pollution 114

Paavola, J. *et al* 2006 209–10
Pakistan 129, 135
Pan African Climate Justice Alliance 180,
    b181, 184
parks: definition *169*; Fairmount Park,
    Philadelphia 161–2; Harlem Piers
    water-front park, New York 174–5, *175*;
    London, Canada 171; Los Angles,
    USA 170–1; model of park use *162*
participation, interrelations with
    distribution and recognition *65*
participatory methods, data gathering 62–3
Patel, Z. 31
Peek, Bobby 30
Pelling, M. 135, *136*
Pellow, D.N. 69–70, 74, 97, 98, 100
Pettit, J. 179
Phillips, C. V. and Sexton, K. 11

pluralism 11, 44–5, 65, *65*, 69–70, b157
political agendas: claim-making 40;
  environmental justice 219–20;
  flooding 130; greenhouse gas emissions
  183, 185–6; research designs 90
political ecology *see also* urban political
  ecology 71–2, *71*, 128, 130, 135
polluter-industrial complex 19–20, 70
polluter pays principle b45, 46–7,
  106, 203
polychlorinated biphenyl (PCB) 78, 79–80
poverty: and air quality 106–7, **106**, 109,
  111; and flood risk 136–9; population
  affected by Hurricane Katrina
  141–2, *141*
power relations: political 185–6; social
  72–4
preferred space 52–3, 151
Principal of Prima Facie Political
  Equality 47
procedural justice: adaptation policies
  196–8; air quality **106**, 107, 123–5;
  climate justice 195–8; community
  consent 98–100; definition 10, b10;
  evidence 217; flood mitigation 153–4;
  grassroots civil rights activism 123–5;
  greenspace planning 176–7; Hurricane
  Katrina **148**; interrelation with other
  concepts of justice *65*; lack of in South
  Africa 31; as part of environmental
  justice framework 47–50, 218–19; waste
  site locations 82, **83**, 99, 101
process: and claim making 40–1, *40*, 64–6;
  interrelations with participation and
  recognition *65*; process explanations 74;
  waste site locations **91**
protest actions *see also* civil rights
  movement (USA): church groups 79;
  multi-scalar nature 83–4
proxy data 56–7, 183
public health expertise 117
Pulido, L. 67, 92, 101, 114–15

qualitative analytical methods 61–3
quantitative analytical methods 55–61, 84,
  88, 107–8

race *see also* environmental racism: access
  to parks 170; as factor within environmental justice in USA 20; flood risk
  131–3, *132*; impacts of climate change
  190–1; lack of in UK environmental
  justice 26; profile of fatalities, Hurricane

Katrina **143**; waste site locations 84–90,
  b85, b87–8, *88*, 91–4, **91**
Rawls, John 44, 194
Real World Coalition 25
recognition, justice as *see also*
  misrecognition: cultural 50–1, 79–82,
  83, 139, 195–8, 218–19; flood
  mitigation 153–4; interrelations with
  distribution and participation *65*
research designs b58–9
risk assessments, cumulative 59–61, b60
rivers: flooding 133, *133*, 134, 135;
  management b45; water quality b58–9
Roberts, J. 188, 193
Roberts, J. and Parks, B.C. 200, **201**,
  203, 204
Roma (people) 24, 139

Sayer, A. 186
scales of analysis 17, 22–3, 35–6, 184–5,
  190, 215–16
Schlosberg, D.: capabilities framework 53;
  distributive justice 42; interrelation of
  concepts of justice 11, 35, 36–7, 51–2,
  65, 218; procedural justice 48;
  recognition, cultural 50
Schnaiberg, A. 68–9
scientific approaches: air quality standards
  122, 123, 125; climate change 182–4;
  doubts over 41, 54, 61
Scotland: air quality studies 111;
  environmental justice as policy
  objective 28–30, 33; Friends of the
  Earth Scotland (FoES) 28–9, *29*, 30, 34;
  Glasgow, proximity to greenspaces **165**,
  168–9; Greengairs 79–80, **83**, 84, 95;
  greenspace use by children b164; waste
  site locations b93–4; woodlands 166–7,
  167, *167*, *168*
Scottish Environmental Protection Agency
  (SEPA) 80
Scottish National Party 29–30
self-determination **195**, 196–7
Sen, A. 52–3, 151
Sharkey, P. 140, 143–4, **143**, 146
Sheppard, E. *et al* 1999 **110**, 111
Shrader-Frechette, K. 47, 54, 55, 63, 64
small island states 195–6, 196–7
Smith, Neil 72
social difference, in environmental terms
  2, 215–16
social movements *see also* grassroots civil
  rights activism 4, 17–18, 24

socio-economic factors: flood risk 133–5; UK research 26
socio-political processes 69–70
socio-spatial analyses: exposure to air pollution 107, 114, 116–17; greenspaces 170–2; urban greenspaces 165–6
South Africa: anti-apartheid movement 31; compared with United Kingdom 32–3; Durban 30–1, 32; emergence of environmental justice movement 30–2, 33; environmental racism 31, 33, 67; links with the USA 30, 31, 36; sustainable development 37
South African Exchange Programme on Environmental Justice (SAEPEJ) 30
South America 35
South Durban Community Environmental Alliance (SDCEA) 30–1, 32
spatial analyses 130, 190, 216
stakeholders 69–70
state, role of 130, 149
statistical analysis: methodologies 216–18; methodologies, air quality 107–11, *108*; qualitative analyses 61–3; quantitative analyses 55–61, 84, 88; siting of waste locations 85–8
Steger, T. b8–9
Stephens, C. *et al* 2001 b8–9, 10
stream rehabilitation, Australia b45
structural process claims 64–5, 219–20
sustainable development 28, 31, 33, 34, 37, 101
Sustainable South Bronx 175–6

Tai-Power 80–2
Taiwan 34, 80–4, *81*, **83**, 99
Tampa, Florida 172
Tavalu (people) 197
technology 68–9
think-tanks 63–4
Tol, R.S.J. *et al* 2003 188
Tompkins, E.L. 194, b194
toxic contamination, flood water 146
toxic imperialism 95–8
toxic waste 79
trade agreements 35, 95
Trafigura b96–7
Transatlantic Initiative on Environmental Justice 24
transnational networks 24, 34–5, 97–8
transport emissions per person *205*

treadmill of production 68–9
triple jeopardy for health 116–17, 122, 126
Tuvalu 195–6

Ueland, J. and Warf, B. 131–3, *132*
UK Black Environmental Network (BEN) 25
United Church of Christ 56, 66, 85, b85, 87
United Kingdom *see also* Scotland: air quality studies 111–14, 120–1; community volunteering 99; compared with South Africa 32–3; emergence of environmental justice movement 25–30, 33, 37; Environmental Justice Summits 28; evidence production 54–5; flooding 127, 133–4, 136–8, 194; framing, characteristics of 25–30, b27, 32–3; government policies 25, 26–8, b27; impacts of climate change 190; nitrogen dioxide levels *112*, *113*, *121*; traffic pollution 118–19; travel-related emissions b205–6, *205*; waste site locations b93–4
United Nations (UN) 194, *199*, *200*
United States Environmental Protection Agency (EPA) b8–9, 10, 18, b19, 23
United States of America *see also* Hurricane Katrina: air quality studies 109–11, **110**, 117–18; Alaska 196, 197; carbon mitigation 201–2; climate change impacts by race 190–1, b191–2; emergence of environmental justice movement 2, 17–23, 34, 56; Environmental Justice Summits 28; environmental racism 2, 18, 66–7, 86, 90–4, 109–11; flood risk and race 131–3, *132*; framing, characteristics of 20–3, b21–2; Gary, Indiana 114, b115; government policies 18–20, b19, 23; grassroots civil rights activism 18, 20, 28; Houston, City of 89; Indianapolis, Indiana 171–2; Just Climate Change Policies b211–12; links with South Africa 30, 31, 36; Los Angeles 170; New York 174–5, *175*, 214; parks 161–2, 170, 171; siting of waste facilities 78–9, 82–90, **83**, *88*, **90**; Tampa, Florida 172; Warren County, North Carolina 78–9, 82–3, **83**, 84, 85
units of analysis 180

urban greenspaces: children's access
to 161–2, b164, 170, 171; community garden, New York (photo) *159*;
community participation 159, 174–5,
176–7; community-scale compensation
175–6; conflicts over use 44–5, 162–3;
cross-cutting themes b157; distributive
justice 176–7; within an environmental
justice framework 156–7; evaluation of
160–3; food justice 159–60; food justice
(photo) *159*; health benefits 158;
London (photo) *161*; model of park use
162; as moderators of environmental
risks 160; parks, USA 161–2, 170–1,
174–5, *175*; proximity to, Glasgow
168–9; statistical methodologies 165–6,
170–2; types of greenspaces 165–6, **169**;
urban greenways 171–2; users 163–5,
b164; value of 157–60
urban political ecology 72–4, b73, 113

vulnerability: age 4, 46, 143–4, **143**; air
quality 106–7, **106**, 115–18, 126; to
environmental impacts 46, 216; to
flood risk 127–8, 135–9, *136*, 147, 149;
impacts of climate change 187

Wales: air quality studies 111, 114, 120–1;
nitrogen dioxide levels *113*; waste site
locations b93–4
Walker, G. *et al* 2003 *112*, *113*, *121*
Walker, G. *et al* 2007 120–1, *121*, 133,
133, *134*, 136–8
Warren County, North Carolina, USA
78–9, 82–3, **83**, 84, 85
waste site locations: analytical themes
b78; criteria of siting process **49**;
as a desirable commodity 98–100;
displacement of waste products 94–5;

distributive justice 82, **83**; emergence
of environmental justice movement
18; environmental racism 84–90, b85,
b87–8, *88*, 91–4, **91**; examples of
claim-making b6; Greengairs, Scotland
79–80, 82–4, **83**; host community, racial
make-up 88–90, *88*; methodologies,
statistical 85–8; metric of distribution
43; Orchid Island (Lanyu), Taiwan
80–4, *81*, **83**, 99; prevention through
reduction 100–1; procedural justice 82,
83, 99, 101; statistical studies, USA
84–8, b87–8, **90**; Warren County, North
Carolina, USA 78–9, 82–4, **83**
Watts, M. and Peet, R. *71*
weather events 184
Weinberg, Adam S. 64
Wenz, P.S. 11, 12–13
West Harlem Environmental Action
(WE ACT): air quality campaigners
(photo) *105*; air quality campaigns
123–5, *124*; water-front park 174–7;
water-front park (photo) *175*
Wheeler, David 96
Wolch, J. *et al* 2002 170–1, 174
women: misrecognition of evidence 62,
63, 68; role, post Hurricane Katrina
153–4; and urban greenspaces
161–2
woodlands: forestry projects 207;
mitigation of environmental risk 161;
Scotland 167, *167*, *168*
World Bank 95
World Disasters Report (International
Federation of Red Cross and Red
Crescent Societies 2007) 139

Yami (people) 80–4
Young, I.M. 147, 221